Polymer Nanocomposites for Dielectrics

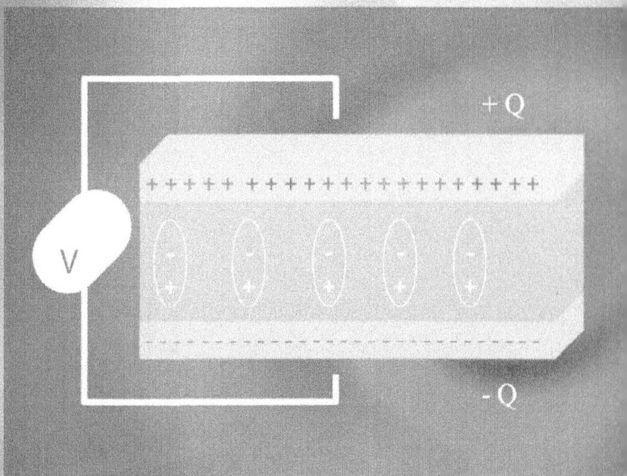

Polymer Nanocomposites for Dielectrics

edited by

Wei-Hong Zhong
Bin Li

Pan Stanford Publishing

Published by

Pan Stanford Publishing Pte. Ltd.
Penthouse Level, Suntec Tower 3
8 Temasek Boulevard
Singapore 038988

Email: editorial@panstanford.com
Web: www.panstanford.com

British Library Cataloguing-in-Publication Data
A catalogue record for this book is available from the British Library.

Polymer Nanocomposites for Dielectrics

ISBN 978-981-4745-53-6 (Hardcover)
ISBN 978-981-4745-54-3 (eBook)

Printed in the USA

Contents

Preface ix

1. Basics of Polymers and Polymer Nanocomposites as Dielectric Materials **1**

Bin Li and Wei-Hong Zhong

1.1 Polymers as Dielectric Materials 1

1.1.1 Structures of Polymers 2

1.1.2 Properties of Polymer Materials 3

1.1.3 Applications of Polymers as Dielectric Films 4

1.2 Polymer Nanocomposites 5

1.2.1 Nanomaterials 5

1.2.2 Dispersion of Nanomaterials in Nanocomposites 6

1.2.3 Interfaces of Polymer Nanocomposites 7

1.2.4 Dielectric Properties of Polymer Nanocomposites 9

2. Theoretical Analysis of Dielectric Relaxation in Polymer Nanocomposites **17**

Bin Li and Wei-Hong Zhong

2.1 Polarizations and Dielectric Properties 17

2.1.1 Dipoles and Dipole Moment 17

2.1.2 Polarization Mechanisms 18

2.1.3 Capacitance 21

2.2 Frequency-Dependent Dielectric Properties 22

2.3 Dielectric Relaxation Models 24

2.3.1 Dielectric Relaxation Models for Polymeric Materials 24

2.3.2 Models for Polymer Nanocomposites 25

2.4 High-Field Dielectric Relaxation 28

3. Perovskite Ceramics and Their Polymer Nanocomposites 35

Bin Li and Wei-Hong Zhong

3.1 Crystal Structures of Barium Titanate 36
3.2 Barium Titanate/Polymer Nanocomposites 39
3.3 Surface Modification of $BaTiO_3$ in Polymer Nanocomposites 42
3.3.1 Surface Modification via Organic Molecules 43
3.3.2 Surface Modification via Polymers 48

4. Graphitic Nanomaterials and Their Polymer Nanocomposites for Dielectric Applications 69

Bin Li and Wei-Hong Zhong

4.1 Graphitic Carbon Nanomaterials 71
4.2 Percolation Phenomenon in Graphitic Nanomaterials/Polymer Nanocomposites 72
4.3 Dispersion and Distribution of Graphitic Nanomaterials 74
4.4 Surface Modification of Nanomaterials 83

5. Poly(Vinylidene Fluoride) in Dielectric Nanocomposites 95

Bin Li and Wei-Hong Zhong

5.1 Crystal Structures of PVDF 96
5.2 Effects of Processing on PVDF Crystal Structures 99
5.3 Effects of Nanomaterials on Phase Transformation in PVDF 100
5.4 Effects of Surface Properties of Nanomaterials on Phase Transformation in PVDF 107

6. Copolymers in Dielectric Polymer Nanocomposites 119

Bin Li and Wei-Hong Zhong

6.1 Polarization Mechanisms in PVDF and PVDF Copolymers 121
6.2 Dicopolymers of PVDF and Their Nanocomposites 123
6.3 Tercopolymers of PVDF 131

7. Dielectric Properties of Bionanocomposites **139**
Guan Gong and Bin Li
7.1 Introduction 140
7.2 Dielectric Properties of Bio-Assembled Nanocomposites 141
7.3 Dielectric Properties of Nanocomposites with Bio-Based Components 145
7.3.1 Biofibril-Based Nanocomposites 145
7.3.1.1 Natural fiber-based nanocomposites 145
7.3.2 Cellulose Nanocrystal-Based Nanocomposites 146
7.3.3 Carbon Nanomaterial-Based Nanocomposites 151
7.3.4 Ceramic Oxide-Based Nanocomposites 155
7.3.4.1 "Traditional" ceramic-based nanocomposites 157
7.3.4.2 "Advanced" ceramics-based nanocomposites 158
7.3.5 Precious Metal-Based Nanocomposites 160
7.4 Summary 162

8. Hybrid Polymer Nanocomposite Systems **171**
Bin Li and Wei-Hong Zhong
8.1 Hybrid Systems with Isolated Nanomaterials 172
8.2 Covalently Bonded Hybrid Nanomaterials 175
8.3 Core–Shell Hybrid Nanostructures 179
8.4 Composite Nanomaterials 184

Index 193

Preface

Numerous books on polymer nanocomposites are available today to readers having different backgrounds. But what caught our attention was the lack of books that exclusively address the dielectric properties of polymer nanocomposites, in spite of being a highly active research area. Besides playing an essential role in electronics and energy applications, the knowledge of dielectric properties provides an efficient and powerful analytical tool for studying polymer nanocomposites. Therefore, we decided to write this introductory book on dielectric properties of polymer nanocomposites with a special coverage of electrical energy storage. It has been written to provide a useful reference material for scientists, engineers, as well as everyone who needs basic understanding on this topic.

The objective of this book is to address fundamental issues in dielectric polymer nanocomposites, as well as strategies to improve the dielectric performances of polymer nanocomposites. In particular, it focuses mostly on the research published over the past 10 years to provide an up-to-date and relevant knowledge and information. It provides a brief introduction of polymer materials and polymer nanocomposites in Chapter 1, which emphasizes the common issues that are critical to dielectric properties of polymeric materials, including structures and properties of polymer matrix and dispersion of nanomaterials and interfaces. Chapter 2 introduces basic theories and models frequently used in theoretical analysis of dielectric properties of polymer nanocomposites. The application of these theories and models on various representative dielectric polymer nanocomposite systems is discussed in detail along with a summary of their limitations in Chapters 3–8. Chapters 3 and 4 focus on two major types of nanomaterials, perovskite ceramic nanomaterials and carbon nanomaterials in dielectric polymer nanocomposites, while Chapters 5 and 6 discuss fluoropolymers and their copolymers in dielectric polymer nanocomposites. As you may notice, we do not intend to cover all dielectric polymer nanocomposites in this book, instead, only some representative dielectric polymer nanocomposites are discussed, which will help the readers gain basic understanding of the roles of polymer matrixes, dispersion issues, and interfaces in dielectric polymer nanocomposites. Chapter 7 extends to contemporary interests in bio-based polymer nanocomposites for dielectric applications, containing either bio-based polymer materials or bio-based nanomaterials, or both, while Chapter 8 introduces a more complex nanocomposite structure: hybrid polymer nanocomposites, which suggest an alternate route to desirable dielectric properties in polymer nanocomposites.

Lastly, we would like to thank Pan Stanford Publishing Pte. Ltd. for its support and excellent job on this book. We are particularly grateful to Stanford Chong and Shivani Sharma. We are also deeply thankful to Dr. Guan Gong (Swerea SICOMP AB, Sweden) for her valuable contribution in Chapter 7. Dr. Zhong would like to acknowledge the partial support from the United States National Science Foundation via the grant NSF CMMI 1463616.

Wei-Hong Zhong

Bin Li

Fall 2016

Chapter 1

Basics of Polymers and Polymer Nanocomposites as Dielectric Materials

Bin Li[a] **and Wei-Hong Zhong**[b]
[a]*Department of Mechanical Engineering, Wichita State University, 1845 Fairmount Street, Wichita, Kansas 67260-0133, USA*
[b]*School of Mechanical and Materials Engineering, Washington State University, 405 NE Spokane Street, Pullman, Washington 99164-2920, USA*
bin.li@wichita.edu, katie_zhong@wsu.edu

1.1 Polymers as Dielectric Materials

Polymers, or macromolecules, have existed in nature probably as long as metals and ceramic materials. However, the massive commercialization and applications of polymer materials have only been around nearly a century, starting with the milestone publication by Staudinger in 1920s, in which the concept of "macromolecule" was proposed as long chain of repeating small molecular unites linked by covalent bonds [1], as well as the invention of Nylon and other important synthetic polymer materials in early 20th century. Due to the unique structures and properties, in contrast to metals and ceramics, polymer materials have been playing increasingly important roles in almost every aspect of our lives, such as transportation, electronics, energy systems, as well as biomedical applications.

Polymer Nanocomposites for Dielectrics
Edited by Wei-Hong Zhong and Bin Li

ISBN 978-981-4745-53-6 (Hardcover), 978-981-4745-54-3 (eBook)
www.panstanford.com

1.1.1 Structures of Polymers

The molecules of polymers are usually composed of thousands of covalently bonded repeating structural units, therefore, the polymer materials have very high molecular weight, compared with metals and ceramics. Typically, the molecular weight of polymer materials is at least 10,000 g/mol. For some polymer materials, such as ultra-high molecular weight polyethylene (UHMWPE), their molecular weight can be up to several million g/mol.

The high molecular weight is considered as one of the most important structural factors in polymers, because of its significant impacts on a couple of properties and the supramolecular structures of polymer materials. In general, the high molecular weight can lead to high mechanical performances, as a result of strong inter-/intra-molecular entanglement. Meanwhile, it is also responsible for the increased viscosity of polymer melts and solutions, which are critical to polymer processing and crystallization of semi-crystalline polymers. Besides high molecular weight, the molecular weight of polymers often shows broad distribution, suggesting different lengths of polymer chain structures caused by various polymerization mechanisms and catalysts. The molecular weight distribution also plays significant roles in polymer structures, mechanical, and rheological properties. However, both high molecular weight and broad molecular weight distribution have very limited effects on physical properties of polymer materials, such as electrical and dielectric properties.

When the polymer has a flexible chain structure, this polymer could develop into highly ordered crystal structures. However, for polymer materials, it is very difficult to achieve 100% crystallinity. Thus, semi-crystalline polymer is usually used to name this group of polymer materials. Several factors hinder the formation of crystal phase in polymers, including double/triple bonds, bulky groups (such as phenyl groups), highly polar molecular structures, etc. These structures are able to stiffen the molecular chain structures and create obstacles for polymer chains to fold into crystal lamellae in polymer materials.

The effects of crystal structures on material properties are more complicated than that of molecular weight. Factors including crystallinity, crystal phase and morphologies are all contributing factors to the properties of polymer materials. In general, the increase in crystallinity could improve the strength and modulus of polymer materials, and enhance barrier properties, while ductility is reduced. Meanwhile, different crystal phases also possess different mechanical performances. For example, β-crystal in polypropylene is considered as a tougher phase compared with α-crystal in polypropylene, thus many attempts have been made to grow more β-crystal in polypropylene for better toughness.

Regarding dielectric performances, the effects of crystal structures are more significant. Poly(vinylidene fluoride) (PVDF) is the best example. PVDF has several different crystal phases (α, β, γ, δ) showing different polarizability in electric field. While it is the most common and stable crystal phase in PVDF, α-phase is less polar compared with other crystal phases. Thus, in terms of dielectric and piezoelectric performances, α-phase is not as desirable as other polar phases. The creation of different crystal phases could be achieved via controlling processing conditions, such as stretching and recrystallization [2–8], as well as using nanomaterials. More discussion on PVDF will be done in Chapter 5.

1.1.2 Properties of Polymer Materials

The growing popularity of polymer materials is due to many unique properties compared with metals and ceramics. Lightweight is probably one of the most attractive properties of polymer materials. Most polymer materials consist of only C, H, O, and N, all of which are lightweight elements. Thus, although the polymers have large molecular structures, the mass density of polymer materials is very low. Typically, their density is close to 1 g/cm^3. This is very important to the development of lightweight vehicles with high fuel efficiency as well as portable consumable electronics. Other advantages of polymer materials also include ease of processing, tunable flexibility, excellent electrical insulation performances, etc. Thus, polymer materials are getting increasing attention in many industrial sectors that are dominated by metals and ceramics.

Based on chemical structures, fabrication techniques, as well as mechanical performances, polymer materials are generally classified into three categories: thermoplastics, thermosets, and elastomers. Thermoplastics usually have linear or branching chain structures, which could be melted or dissolved in solvents, therefore, the processing of this type of polymer materials could be conducted in both molten state and solutions. Also, the resulting products could be re-processed and recycled. In thermoplastics, the secondary bonding plays a critical role in their mechanical and thermal properties, for example, usually low thermal resistance compared with that of thermosets with 3D cross-linking structures. The 3D network structures bring excellent mechanical performances and thermal stability to thermoset plastics, thus, thermoset plastics are very popular in construction of light weight structures for long term applications. Elastomers show non-linear elastic deformation within a broad range of strain. They possess excellent damping and sealing performances compared with other classes of polymers. Furthermore, due to their outstanding capability of recoverable deformation, elastomers have also been frequently studied to develop new smart materials with high sensitivity to external stimuli [9–15].

1.1.3 Applications of Polymers as Dielectric Films

Most of the polymer materials are excellent electrical insulators showing very high electrical resistance and breakdown strength, leading to applications as electrical insulations. The electrical insulating materials are also called dielectric materials, ideal for use in capacitors for various electronics and energy devices including electronic circuits, power electronic converters, inverters, etc. In addition to light weight, there are a few more advantages using polymer dielectric films, including compact size, high emissivity, low absorption, great potential to complex 3D design of new generation electronics, and so on. Currently, the popular commercial polymer dielectric materials are polypropylene (PP), polyethylene terephthalate (PET), polyphenylene sulfide (PPS), polyethylene naphthalate (PEN), polytetrafluoroethylene (Teflon) (PTFE), poly(vinylidene fluoride) (PVDF), etc. Among them, BOPP and PET are dominating polymer dielectric films today, due to their high breakdown strength, less dielectric loss, and so on [16].

The disadvantage of polymer materials as dielectrics is also obvious: compared with dielectric ceramics with a dielectric constant as high as a few thousands, low dielectric constant is the primary issue for polymer dielectric materials. A typical dielectric constant of polymer materials is between 2~10. Thus, how to improve the dielectric properties has been an essential research topic in development of high-performance polymer dielectric films. One effective approach is fabrication of polymer nanocomposites.

1.2 Polymer Nanocomposites

Hybridizing polymer materials and functional nanomaterials is believed to be an effective approach to achieving enhanced performances and creating new functionalities for the light-weight material systems. The resulting polymer nanocomposites are expected to possess advantages from both polymer matrix and the nano-reinforcement. In past few decades, polymer nanocomposite is one of the most attractive research areas drawing magnificent attentions from academia as well as industry. It has been reported that the addition of nanomaterials to polymer matrices could improve mechanical and thermal properties [4, 17–24], and/or increase electrical conduction, dielectric constant, etc. [25–31].

1.2.1 Nanomaterials

The prosperous nanotechnology has created a great number of nanomaterials and nanostructures available for making various functional polymer nanocomposites. Besides downsizing bulk materials to nanoscale, many new nanostructures have been created, such as nanofibers, nanotubes, nanoneedles, and fullerene [32–35]. According to International System of Unit, nanomaterials have at least one dimension smaller than 100 nm. Compared with micro-size materials and bulk materials, the nanomaterials have much larger surface areas and more active surface atoms, leading to many novel optical, electrical, and magnetic phenomena that could only be understood by quantum mechanics. These novel phenomena provide great opportunity to realize new functions in polymer nanocomposites.

In terms of improving dielectric properties for polymer materials, graphitic carbon nanomaterials (i.e., carbon nanofiber, carbon nanotube, graphene, etc.) and dielectric ceramic nanomaterials (i.e.,

TiO_2, ZnO, $BaTiO_3$, etc.) have proven to be the most effective fillers [36–40].

1.2.2 Dispersion of Nanomaterials in Nanocomposites

Besides the intrinsic structures and physical properties, dispersion of a nanomaterial is considered as the most critical factor affecting the properties and performances of the polymer nanocomposites including dielectric properties. In other words, to achieve high-performance polymer nanocomposites, an effective control of nanomaterial dispersion is the prerequisite. Figure 1.1 lists four typical dispersion and distribution states of a nanomaterial in a polymer matrix: (a) uniform dispersion and uniform distribution; (b) uniform dispersion and poor distribution; (c) poor dispersion and uniform distribution; and (d) poor dispersion and poor distribution. Taking electrical conduction as an example, a conductive pathway can be more effectively formed when the conductive nanomaterials show the uniform dispersion and poor distribution state as shown in Fig. 1.1 (b). In a previous study [41], the negative dielectric constant was found when the carbon nanofibers showed similar dispersion and distribution states to Fig. 1.1 (c). To achieve a high dielectric constant in the polymer nanocomposites, it is believed that both uniform dispersion and uniform distribution of nanomaterials should be realized in the polymer matrix [Fig. 1.1 (a)]. Furthermore, the uniform dispersion and distribution have also shown positive contributions to the enhancement in mechanical and thermal properties of the polymer nanocomposites [42–46].

Therefore, pursuit of uniform dispersion of nanomaterials is very essential to fabrication of high-quality polymer nanocomposites. However, as mentioned earlier, nanomaterials have high surface area and a great number of active surface atoms, which lead to high surface energy for the nanomaterials. To minimize the high surface energy, the nanomaterials tend to aggregate together and form agglomerates in the polymer matrix. Great efforts have been made to achieve uniform dispersion, including mechanical shearing, ultrasonication, chemical modification, in-situ polymerization, etc. [23, 47–50]. In most successful cases, more than one of these approaches were applied to assure the uniform dispersion. For

1D nanostructures, such as nanofibers and nanotubes, in addition to the high surface energy, the strong entanglement also prevents formation of uniform dispersion of nanomaterials, which makes improving dispersion uniformity more challenging.

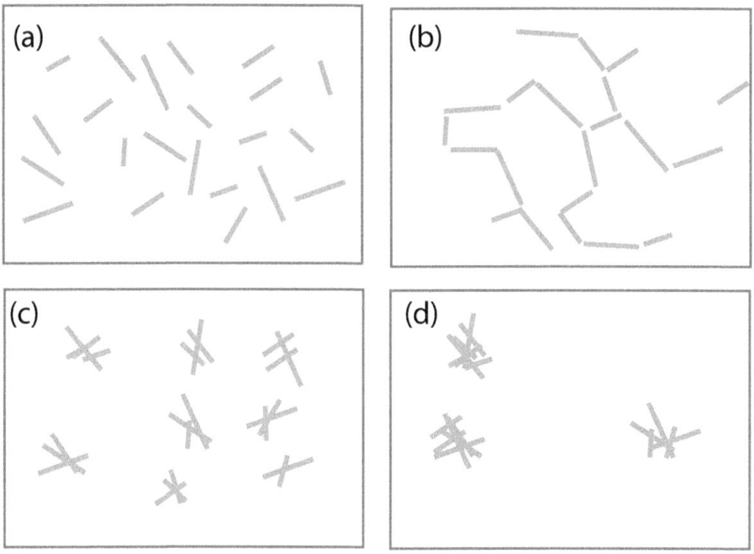

Figure 1.1 Schematic sketches of four dispersion and distribution states of a 1D nanostructure in polymer nanocomposite.

1.2.3 Interfaces of Polymer Nanocomposites

Interface is an important structure in polymer nanocomposites and other multiple-phase material structures. In this chapter, the interface issue is brought up, because of two reasons. First, it is related to the dispersion of nanomaterials, and second, the interfacial property, that is, interfacial polarization phenomena, is critical to the dielectric properties of polymer nanocomposites.

Regarding its relationship with dispersion of a nanomaterial, the incompatible surface characteristic of the nanomaterial with polymer matrices is responsible for the poor dispersion of the nanomaterial in polymer matrices. Thus, modifications of both nanomaterial and polymer matrix have often been applied to improve the interfacial compatibility between polymers and nanomaterials, which can lead to improved dispersion. The resulting nanocomposites usually showed enhanced mechanical, thermal, and electrical properties (when conductive nanomaterials were used). Meanwhile, interfacial polarization is the primary polarization mechanism in the frequency range for many capacitor applications. Thus, the structures and electronic properties at the interfaces are very important to the dielectric properties of the resulting nanocomposites. However, the knowledge in this area is still very limited. Fig. 1.2 gives an example of the contribution of interface between $BaTiO_3$ (BT) nanoparticles and PVDF matrix, or surface modification of $BaTiO_3$ nanoparticles, to the electrical breakdown strength of BT/PVDF nanocomposites. The treated BT nanoparticles were obtained by modifying pristine BT nanoparticles with tetrafluorophthalic acid. The resulting nanocomposites showed enhanced electrical insulation performance, which is directly related to the surface modification by tetrafluorophthalic acid. The surface modification significantly improved the dispersion of the BT nanoparticles by forming a robust binding between BT nanoparticles and PVDF matrix. At the same time, the surfactant could form a passive layer to reduce the generation of charge carriers that will lead to electrical conduction and lower breakdown strength [1] .

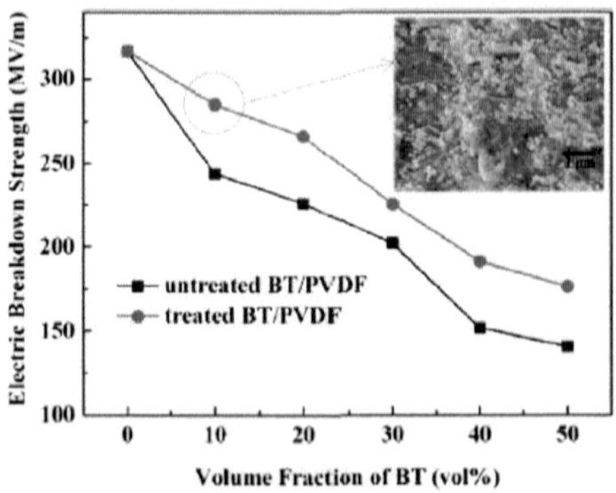

Figure 1.2 Electric breakdown strength of the pure polymer matrix and $BaTiO_3$ (BT)/PVDF nanocomposites as a function of the volume fraction of BT nanoparticles. (The inset is the SEM image of the cross surface of the nanocomposite thin films with 10 vol.% surface treated BT nanoparticles.) Reprinted with permission from Ref. [1], Copyright 2013, AIP Publishing LLC.

Compared with modifying polymer matrix, surface modification methods for nanomaterials appear to be more efficient. The surface modification could substantially alter the surface properties of the nanomaterials, also individualize and stabilize nanomaterials in a solution or a polymer matrix. Current surface modification can be divided into two categories: covalent modification and non-covalent modification. In covalent surface modification, reactive groups, such as carbonyl groups (–COOH), hydroxyl groups (–OH), and amine groups (–NH_2), are introduced onto the surface of nanomaterials. The chemical reaction between surfactants and these reactive groups could form strong covalent bonds. In non-covalent modifications, the surfactant molecules could directly wrap up the nanomaterials via non-covalent interactions, such as π–π stacking between conjugated structures and graphitic structures [51, 52]. As a result, the dispersion quality could be largely improved. Compared with the covalent surface modification approaches, non-covalent surface modification has only achieved limited applications. Currently, most related reports in polymer nanocomposites have focused on graphitic carbon nanomaterials and their nanocomposites.

1.2.4 Dielectric Properties of Polymer Nanocomposites

Applying functional nanomaterials to polymers has achieved great success in improving dielectric properties of polymers via controlling the dispersion and interfacial structures. Dispersion and interfacial structures can be seen as the most important and critical factors of dielectric properties, in addition to the dielectric properties of both polymers and nanomaterials, therefore, the most effort in this area has been made to achieve uniform dispersion quality and design optimal interfacial structures in polymer nanocomposites. It has been reported that, with addition of a small amount of graphite nanoplatelets in PVDF, the low frequency dielectric constant can go up to $\sim 10^9$, in contrast to ~ 10 for pure PVDF [53]. In this book, Chapters 3–8 will review the similar success in significant polymer nanocomposites, and will also summarize useful knowledge for future optimal design of polymer nanocomposite based dielectric films.

Furthermore, the book will also refer to several important, but less understood or less successful issues in development of polymer nanocomposite based dielectric films.

The addition of electrically conductive nanomaterials in polymers usually results in very high dielectric constant. The highest reported dielectric constant occurred to this group of polymer nanocomposites, suggesting their potential in energy storage application. However, with the dramatic increase in dielectric constant, the electrical conductivity also increases, due to the formation of conductive network. This simultaneous increase in electrical and dielectric properties leads to large current leakage as well as high dielectric loss, which are unwanted phenomena. In particular, the high dielectric loss issue is common to all dielectric materials. High dielectric constant is typically accompanied with high dielectric loss, due to the dielectric relaxation mechanisms. The high loss will not only lower the device efficiency, but also cause so-called dielectric heating that will deteriorate the device performances. How to effectively achieve very high dielectric constant while keeping the current leakage and dielectric loss low is a very challenging issue.

Another important issue is dielectric relaxation behaviors of polymer nanocomposites in high electrical field. The knowledge in this area is very important to polymer nanocomposite in high-power applications [54–56]. However, the dielectric relaxation of polymer nanocomposites in high electric field is not well understood, while there are a great number of studies on the dielectric relaxation of polymer nanocomposites in low electric field.

References

1. Yu, K., Niu, Y. J., Xiang, F., Zhou, Y. C., Bai, Y. Y., and Wang, H. (2013). Enhanced electric breakdown strength and high energy density of barium titanate filled polymer nanocomposites, *J Appl Phys*, **114**, 174107.

2. Ahmad, Z., Prasad, A., and Prasad, K. (2009). A comparative approach to predicting effective dielectric, piezoelectric and elastic properties of PZT/PVDF composites, *Physica B*, **404**, 3637–3644.

3. Alluri, N. R., Saravanakumar, B., and Kim, S. J. (2015). Flexible, hybrid piezoelectric film (BaTi(1-*x*)ZrxO_3)/PVDF nanogenerator as a self-powered fluid velocity sensor, *ACS Appl Mater Interfaces*, **7**, 9831–9840.

4. An, N. L., Liu, S. L., Fang, C. Q., Yu, R. E., Zhou, X., and Cheng, Y. L. (2015). Preparation and properties of beta-phase graphene oxide/PVDF composite films, *J Appl Polym Sci*, **132**.

5. Andrews, C., Lin, Y. R., Tang, H. X., and Sodano, H. A. (2011). Influence of aspect ratio on effective electromechanical coupling of nanocomposites with lead zirconate titanate nanowire inclusion, *J Intel Mat Syst Str*, **22**, 1879–1886.
6. Ansari, S. and Giannelis, E. (2009). Functionalized graphene sheet-poly(vinylidene fluoride) conductive nanocomposites, *J Polym Sci Pol Phys*, **47**, 888–897.
7. Araujo, M. C., Costa, C. M., and Lanceros-Mendez, S. (2014). Evaluation of dielectric models for ceramic/polymer composites: Effect of filler size and concentration, *J Non-Cryst Solids*, **387**, 6–15.
8. Arlt, K. and Wegener, M. (2010). Piezoelectric PZT/PVDF-copolymer 0-3 composites: Aspects on film preparation and electrical poling, *IEEE Trans Dielectr Electr Insul*, **17**, 1178–1184.
9. Chen, T., Qiu, J., Zhu, K., Ji, H., Fan, C., and Chen, Q. (2013). Preparation and dielectric properties of a polyurethane elastomer filled with resol-derived ordered mesoporous carbon, *J Mater Sci: Mater Electron*, **24**, 2013–2018.
10. Dang, Z. M., Xia, B., Yao, S. H., Jiang, M. J., Song, H. T., Zhang, L. Q., and Xie, D. (2009). High-dielectric-permittivity high-elasticity three-component nanocomposites with low percolation threshold and low dielectric loss, *Appl Phys Lett*, **94**, 042902.
11. Fritzsche, J. and Kluppel, M. (2009). Molecular dynamics of silica and organoclay filled XNBR-nanocomposites, *Kgk-Kaut Gummi Kunst*, **62**, 16–22.
12. Iacob, M., Stiubianu, G., Tugui, C., Ursu, L., Ignat, M., Turta, C., and Cazacu, M. (2015). Goethite nanorods as a cheap and effective filler for siloxane nanocomposite elastomers, *RSC Adv*, **5**, 45439–45445.
13. Kumar, V., Hanel, T., Giannini, L., Galimberti, M., and Giese, U. (2014). Graphene/styrene butadiene rubber nanocomposite, *Kgk-Kaut Gummi Kunst*, **67**, 29–36.
14. Mahapatra, S. S., Yadav, S. K., Yoo, H. J., and Cho, J. W. (2011). Highly stretchable, transparent and scalable elastomers with tunable dielectric permittivity, *J Mater Chem*, **21**, 7686–7691.
15. Melvin, G. J. H., Ni, Q. Q., and Natsuki, T. (2014). Fabrication and characterization of polymer-based electroactive nanocomposite actuator, *Microelectron Eng*, **126**, 9–12.
16. Mackey, M., Schuele, D. E., Zhu, L., Flandin, L., Wolak, M. A., Shirk, J. S., Hiltner, A., and Baer, E. (2012). Reduction of dielectric hysteresis in

multilayered films via nanoconfinement, *Macromolecules*, **45**, 1954–1962.

17. Asimakopoulos, I. A., Psarras, G. C., and Zoumpoulakis, L. (2014). Barium titanate/polyester resin nanocomposites: Development, structure-properties relationship and energy storage capability, *Express Polym Lett*, **8**, 692–707.
18. Bicen, M., Kayaman-Apohan, N., Karatas, S., Dumludag, F., and Gungor, A. (2015). The effect of surface modification of zeolite 4A on the physical and electrical properties of copolyimide hybrid films, *Microporous Mesoporous Mater*, **218**, 79–87.
19. Abdelaziz, M. and Abdelrazek, E. M. (2014). The effect of n anographite addition on the physical properties of poly(vinylidene fluoride)/hydroxypropyl cellulose blend, *J Mater Sci: Mater Electron*, **25**, 5481–5490.
20. Abdolmaleki, A. and Bazyar, Z. (2013). Preparation and characterization of poly(benzimidazole-amide)/ZnO nanocomposites using silane coupling agent, *Polym-Plast Technol Eng*, **52**,. 1542–1549.
21. Agoudjil, B., Ibos, L., Candau, Y., and Majeste, J. C. (2008). A comparative analysis of dielectric, rheological and thermophysical behaviour of ethylene vinyl acetate/BaTiO(3) composites, *J Phys D Appl Phys*, **41**, 055407.
22. Ahmetli, G., Dag, M., Deveci, H., and Kurbanli, R. (2012). Recycling studies of marble processing waste: Composites based on commercial epoxy resin, *J Appl Polym Sci*, **125**, 24–30.
23. Al-Hussaini, A. S., Zoromba, M. S., and El-Ghamaz, N. A. (2013). In situ polymerization and characterization of aniline and O-anthranilic acid copolymer/pyrogenic silica nanocomposites, *Polym-Plast Technol*, **52**, 1089–1096.
24. Ali, M. and Choudhry, M. A. (2015). Preparation and characterization of EPDM-silica nano/micro composites for high voltage insulation applications, *Mater Sci Poland*, **33**, 213–219.
25. Fleming, R. J., Ammala, A., Casey, P. S., and Lang, S. B. (2011). Conductivity and space charge in LDPE/$BaSrTiO_3$ nanocomposites, *IEEE Trans Dielectr Electr Insul*, **18**, 15–23.
26. Frechette, M. F. (2010). Nanodielectrics: A panacea for solving all electrical insulation problems? 2010 International Conference on Solid Dielectrics, Potsdam, Germany, July 4–9, 2010.

27. Fredin, L. A., Li, Z., Lanagan, M. T., Ratner, M. A., and Marks, T. J. (2013). Sustainable high capacitance at high frequencies: Metallic aluminum–polypropylene nanocomposites, *Acs Nano*, **7**, 396–407.

28. Fritzsche, J., Lorenz, H., and Kluppel, M. (2009). CNT based elastomer-hybrid-nanocomposites with promising mechanical and electrical properties, *Macromol Mater Eng*, **294**, 551–560.

29. Frubing, P., Wang, F. P., and Wagener, M. (2012). Relaxation processes and structural transitions in stretched films of poly(vinylidene fluoride) and its copolymer with hexafluoropropylene, *Appl Phys A Mater Sci Process*, **107**, 603–611.

30. Fu, J.-F., Shi, L.-Y., Zhong, Q.-D., Chen, Y., and Chen, L.-Y. (2011). Thermally conductive and electrically insulative nanocomposites based on hyperbranched epoxy and nano-Al_2O_3 particles modified epoxy resin, *Polym Adv Technol*, **22**, 1032–1041.

31. Fu, Y., Liu, L. S., and Zhang, J. W. (2014). Manipulating dispersion and distribution of graphene in PLA through novel interface engineering for improved conductive properties, *ACS Appl Mater Interfaces*, **6**, 14069–14075.

32. Berry, K. R., Russell, A. G., Blake, A., and Roper, D. K. (2012). Gold nanoparticles reduced in situ and dispersed in polymer thin films: optical and thermal properties, *Nanotechnology*, **23**, 375703.

33. Chopra, S. and Alam, S. (2013). Fullerene containing polyurethane nanocomposites for microwave applications, *J Appl Polym Sci*, **128**, 2012–2019.

34. Jiang, Z. Y., Zhang, H., Zhang, Z., Murayama, H., and Okamoto, K. (2008). Improved bonding between PAN-based carbon fibers and fullerene-modified epoxy matrix, *Compos Part A: Appl S*, **39**, 1762–1767.

35. Perez-Maqueda, L. A., Dianez, M. J., Gotor, F. J., Sayagues, M. J., Real, C., and Criado, J. M. (2003). Synthesis of needle-like $BaTiO_3$ particles from the thermal decomposition of a citrate precursor under sample controlled reaction temperature conditions, *J Mater Chem*, **13**, 2234–2241.

36. Al Jaafari, A. A. and Ayesh, A. S. (2011). Effect of ZnO nano-particles on the dielectric relaxation behavior and thermal stability of polycarbonate host, *J Thermoplast Compos Mater*, **24**, 837–852.

37. Beltran, H., Maso, N., Cordoncillo, E., and West, A. R. (2007). Nanocomposite ceramics based on La-doped $BaTi_2O_5$ and $BaTiO_3$ with high temperature-independent permittivity and low dielectric loss, *J Electroceram*, **18**, 277–282.

38. Bouropoulos, N., Psarras, G. C., Moustakas, N., Chrissanthopoulos, A., and Baskoutas, S. (2008). Optical and dielectric properties of ZnO-PVA nanocomposites, *Phys Status Solidi A*, **205**, 2033–2037.

39. Kuang, X. W., Gao, Q., and Zhu, H. (2013). Effect of calcination temperature of TiO2 on the crystallinity and the permittivity of PVDF-TrFE/TiO2 composites, *J Appl Polym Sci*, **129**, 296–300.

40. Rathod, S. G., Bhajantri, R. F., Ravindrachary, V., Sheela, T., Pujari, P. K., Naik, J., and Poojary, B. (2015). Pressure sensitive dielectric properties of TiO_2 doped PVA/CN-Li nanocomposite, *J Polym Res*, **22**, 1–14.

41. Li, B., Sui, G., and Zhong, W. H. (2009). Single negative metamaterials in unstructured polymer nanocomposites toward selectable and controllable negative permittivity, *Adv Mater*, **21**, 4176–4180.

42. Seckin, T., Koytepe, S., Kivilcim, N., Bahce, E., and Adiguzel, I. (2008). Low dielectric constant polyimide-zirconium nanocomposites with improved thermal properties, *Int J Polym Mater*, **57**, 429–441.

43. Sen, S., Mishra, S. K., Sagar, S., and Das, S. K. (2008). Preparation and characterization of PMN-PT nanocomposite, *Indian J Eng Mater S*, **15**, 111–115.

44. Sencadas, V., Lanceros-Mendez, S., Serra, R. S. I., Balado, A. A., and Ribelles, J. L. G. (2012). Relaxation dynamics of poly(vinylidene fluoride) studied by dynamical mechanical measurements and dielectric spectroscopy, *Eur Phys J E*, **35**.

45. Sengwa, R. J., Choudhary, S., and Sankhla, S. (2009). Dielectric spectroscopy of hydrophilic polymers-montmorillonite clay nanocomposite aqueous colloidal suspension, *Colloid Surf A*, **336**, 79–87.

46. Seveyrat, L., Lebrun, L., Jomaa, M. H., Wongtimnoi, B. K., Liu, Q., Guiffard, B., Masenelli-Varlot, K., Cavaille, J. Y., and Guyomar, D. (2013). Influence of the type of carbon nanocharges on the dielectric, mechanical and electroactive properties of polyurethane composites films, *J Optoelectron Adv M*, **15**, 1346–1351.

47. Anuradha, S. and Prasad, M. V. N. A. (2014). Synthesis, characterization, and X-band permittivity studies of polypyrrole-zirconium composite, *Spectrosc Lett*, **47**, 723–728.

48. Li, G., Yu, S. H., Sun, R., and Lu, D. (2015). Clean and in-situ synthesis of copper-epoxy nanocomposite as a matrix for dielectric composites with improved dielectric performance, *Compos Sci Technol*, **110**, 95–102.

49. Ravikiran, Y. T., Lagare, M. T., Sairam, M., Mallikarjuna, N. N., Sreedhar, B., Manohar, S., MacDiarmid, A. G., and Aminabhavi, T. M. (2006).

Synthesis, characterization and low frequency AC conduction of polyaniline/niobium pentoxide composites, *Synth Met*, **156**, 1139–1147.

50. Shevchenko, V. G., Polschikov, S. V., Nedorezova, P. M., Klyamkina, A. N., Shchegolikhin, A. N., Aladyshev, A. M., and Muradyan, V. E. (2012). In situ polymerized poly(propylene)/graphene nanoplatelets nanocomposites: Dielectric and microwave properties, *Polymer*, **53**, 5330–5335.

51. Gu, H. B., Tadakamalla, S., Zhang, X., Huang, Y. D., Jiang, Y., Colorado, H. A., Luo, Z. P., Wei, S. Y., and Guo, Z. H. (2013). Epoxy resin nanosuspensions and reinforced nanocomposites from polyaniline stabilized multi-walled carbon nanotubes, *J Mater Chem C*, **1**, 729–743.

52. Saini, P., Choudhary, V., Singh, B. P., Mathur, R. B., and Dhawan, S. K. (2011). Enhanced microwave absorption behavior of polyaniline-CNT/polystyrene blend in 12.4–18.0 GHz range, *Synth Met*, **161**, 1522–1526.

53. He, F., Lau, S., Chan, H. L., and Fan, J. (2009). High dielectric permittivity and low percolation threshold in nanocomposites based on poly(vinylidene fluoride) and exfoliated graphite nanoplates, *Adv Mater*, **21**, 710–715.

54. Wu, L. W., Wang, X. H., Gong, H. L., Hao, Y. N., Shen, Z. B., and Li, L. T. (2015). Core-satellite $BaTiO_3$@$SrTiO_3$ assemblies for a local compositionally graded relaxor ferroelectric capacitor with enhanced energy storage density and high energy efficiency, *J Mater Chem C*, **3**, 750–758.

55. O'Connor, K. A., Smith, J., and Curry, R. D. (2009). Dielectric characterization of polymer-ceramic nanocomposites, *2009 IEEE Pulsed Power Conference*, Vols 1 and 2, 336–341.

56. Yao, K., Chen, S. T., Rahimabady, M., Mirshekarloo, M. S., Yu, S. H., Tay, F. E. H., Sritharan, T., and Lu, L. (2011). Nonlinear dielectric thin films for high-power electric storage with energy density comparable with electrochemical supercapacitors, *IEEE Trans Ultrason Ferroelectr Freq Control*, **58**, 1968–1974.

Chapter 2

Theoretical Analysis of Dielectric Relaxation in Polymer Nanocomposites

Bin Li[a] **and Wei-Hong Zhong**[b]
[a]*Department of Mechanical Engineering, Wichita State University, 1845 Fairmount Street, Wichita, Kansas 67260-0133, USA*
[b]*School of Mechanical and Materials Engineering, Washington State University, 405 NE Spokane Street, Pullman, Washington 99164-2920, USA*
bin.li@wichita.edu, katie_zhong@wsu.edu

2.1 Polarizations and Dielectric Properties

2.1.1 Dipoles and Dipole Moment

Dipole is the fundamental concept in dielectric theory. When there is a separation between a positive charge and a negative charge, a dipole is formed, as shown in Fig. 2.1. Generally, there are three types of dipoles classified on the basis of the formation mechanisms: instant dipoles, permanent dipoles, and induced dipoles. Instant dipoles exist in all molecules and atoms. The instant separation between positive charge and negative charge is mainly caused by constant thermal motions within molecules and atoms. In this case, the

Polymer Nanocomposites for Dielectrics
Edited by Wei-Hong Zhong and Bin Li

ISBN 978-981-4745-53-6 (Hardcover), 978-981-4745-54-3 (eBook)
www.panstanford.com

dipoles are usually smaller. In contrast, in some molecules, because of the nature of their chemical bonds, the distribution of electrons within the molecules is not uniform, which results in permanent separation between positive charge center and negative charge center, i.e., permanent dipoles. The permanent dipole occurs when the two bonding elements have large differences in electronegativity, such as ionic bonds and some covalent bonds in polar molecules. Furthermore, the asymmetrical molecular structures and secondary bonding in materials significantly contribute to the formation of a permanent dipole. Usually, permanent dipoles are much bigger than instant dipoles. Dipoles can also be induced when the molecules and atoms are exposed to adjacent polar molecules, as a result of electrostatic interactions.

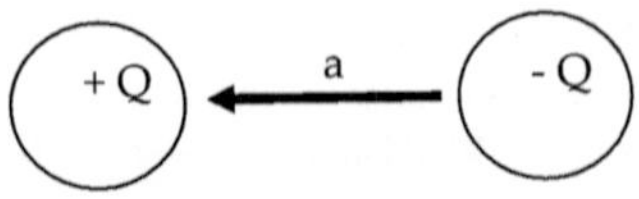

Figure 2.1 Schematic illustration of a dipole.

According to Fig. 2.1, it is obvious that magnitude of a dipole is determined by both the separation distance (**a**) between positive and negative charges and the amount of charges (Q). Thus, a concept of dielectric dipole moment (**p**) is introduced to measure the magnitude of a dipole, as given in Eq. 2.1. Here, **p** is a vector from negative charge to positive charge.

$$\mathbf{p} = Q\mathbf{a} \tag{2.1}$$

2.1.2 Polarization Mechanisms

External electric field could not only induce the formation of dipoles, but also align the permanent and induced dipoles along the direction of the electric field. This process is called polarization. Polarization is responsible for many dielectric phenomena, such as energy storage and piezoelectricity. There are four types of polarization mechanisms: electronic polarization, ionic polarization, orientation polarization, and interfacial polarization. The dielectric materials exhibit at least one of these four polarization mechanisms, depending on both materials structures and external conditions, such as frequency and temperature. Mathematically, polarization (**P**) is defined as the total dipole moment per unit volume, as given in Eq. 2.2:

$$\mathbf{P} = (\mathbf{p}_1 + \mathbf{p}_2 + \mathbf{p}_3 \cdots + \mathbf{p}_n)/\text{Volume} \tag{2.2}$$

Electronic polarization describes the shift of electron clouds within a molecule or an atom under the influence of external electrical field. This shift results in the separation of positive charge center (nuclei) and negative charge center (electron clouds), as shown in Fig. 2.2. It could occur to all the atoms and molecular structures. However, compared with other polarization mechanisms, the dipole moment as a result of electronic polarization is rather small.

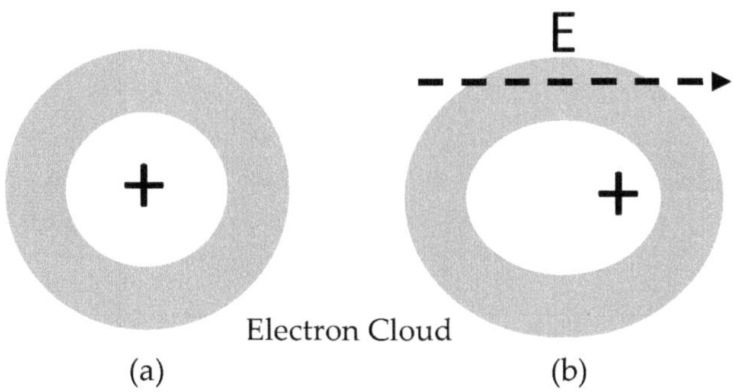

Figure 2.2 Electronic polarization: (a) the electron cloud distribution in the absence of electric field; (b) the shift of electron cloud in electric field.

Ionic polarization only occurs to the materials that have ionic bonds, such as NaCl and $BaTiO_3$. In ionic solids, each pair of neighboring cation and anion can form a dipole and has a dipole moment. However, without external electrical field, the net dipole moment of the solid is zero, because of opposite alignment of adjacent dipoles with same magnitude of dipole moment, as shown in Fig. 2.3. In a AX type ionic solid, such as NaCl, when electric field is applied to the ionic solid, the cations and anions would shift in opposite directions driven by electric field, leading to unequal charge separations ($a' > b$) in adjacent dipoles, as shown in Fig. 2.3. This change induced by external electric field eventually creates a non-zero net dipole moment of the ionic solids.

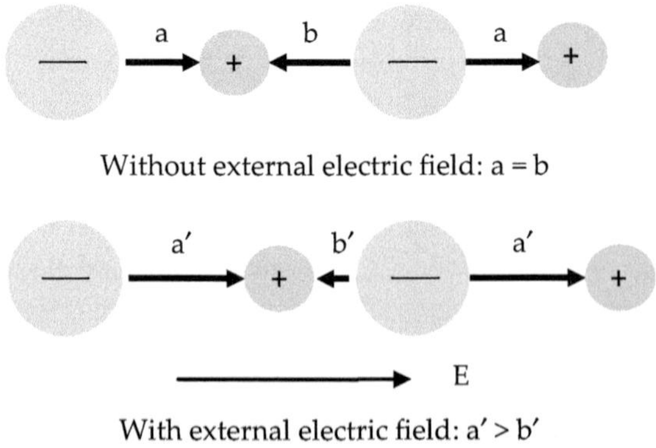

Figure 2.3 Ionic polarization in AX type ionic solid.

Orientational polarization is also called dipolar polarization. Many polar molecules have permanent dipoles. However, the materials consisting of these polar molecules usually do not have net dipole moment because of random orientation of these permanent dipoles in the absence of external electric field, as shown in Fig. 2.4. The randomly oriented dipoles will align along the direction of the applied electric field, leading to a net dipole moment.

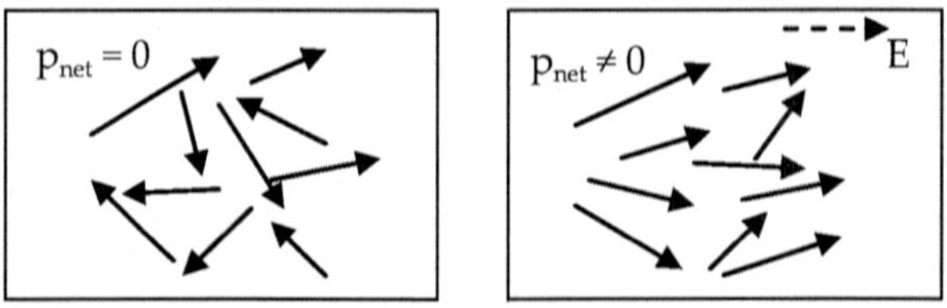

Figure 2.4 Orientational polarization of permanent dipoles (solid arrows) in external electric field.

Interfacial polarization commonly exists in all materials showing multiple phases/regions. Because of the differences in electronic properties, the charge accumulation could be expected at the interface between different phases/regions, such as grain boundaries in polycrystals, the interface between amorphous phase and crystalline phase in semicrystalline polymers, as well as some structural defects and impurities. In polymer nanocomposites, the interface between the polymer matrix and nanomaterials is also able to accumulate charges, which has shown dramatic effects on the dielectric performances of polymer nanocomposites. When the electrical field is applied, the positive and negative charges will migrate accordingly, which leads to the formation of a positive charge layer and a negative charge layer on both sides of the interface, as shown in Fig. 2.5. This interfacial polarization is also prominent between dielectric materials and metallic electrode.

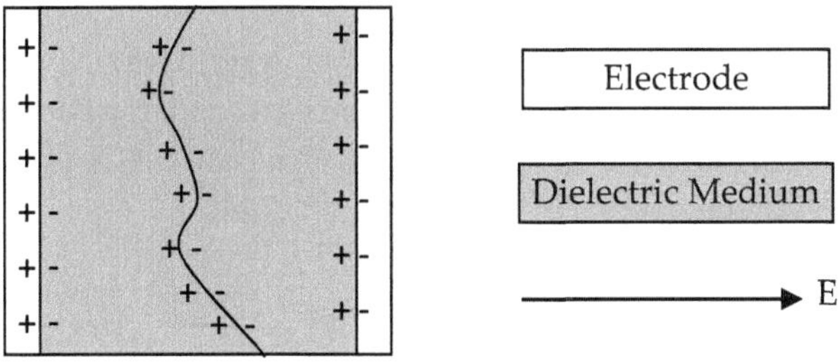

Figure 2.5 Interfacial polarization (charge accumulation) at different interfaces in the presence of external electric field.

2.1.3 Capacitance

When a dielectric material is placed between two metallic plates connected with a constant voltage supply (V), a certain amount of electrical charges (Q) can be stored on both metallic plates, as a result of polarization of dielectric materials. This simple device, as depicted in Fig. 2.6, is the so-called parallel plate capacitor. The amount of stored electrical charges on either plate is described as capacitance, which is given in Eq. 2.3:

$$C = Q/V \tag{2.3}$$

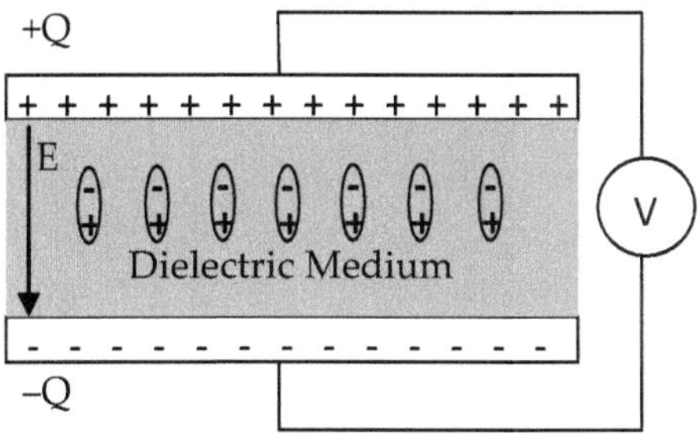

Figure 2.6 A parallel plate capacitor with a dielectric medium between two plates.

The unit of capacitance (C) is coulombs per volt, or farads (F).

Capacitance is a geometry dependent parameter, which can be written as:

$$C = \varepsilon_0 \varepsilon_r \frac{A}{d} \tag{2.4}$$

Where ε_0 the permittivity of vacuum, a universal constant having a value of 8.85×10^{-12} F/m; ε_r is relative permittivity (or dielectric constant) of the dielectric material, which is a material constant; A is the area of the plates; and d is the distance between the two plates, equal to the thickness of the dielectric medium.

2.2 Frequency-Dependent Dielectric Properties

Dielectric performances in materials are resulted from at least one of the four aforementioned polarization mechanisms. The four mechanisms have different relaxation frequencies. Thus, the dielectric properties of materials usually show strong frequency dependence, as shown in Fig. 2.7. To investigate the frequency-dependent dielectric properties, dielectric relaxation spectroscopy is used. Study of dielectric relaxation behaviors could be deployed to measure the dielectric properties at different frequencies, providing useful information for optimal design of materials for dielectric applications. Meanwhile, because different microstructures typically show different relaxation frequencies, similar to dynamic mechanical analysis, dielectric relaxation spectroscopy has been commonly applied in material research [3–9]. The temperature and frequency-dependent dielectric relaxation behaviors are often related to chemical structures, crystal structures and phase structures as well as various materials relaxation behaviors. Therefore, by dielectric analysis, an in-depth understanding of structures and properties of polymeric materials can be obtained.

The most important parameter obtained from dielectric relaxation spectroscopy is dielectric permittivity (ε^*), which is a complex number, as given in Eq. 2.5:

$$\varepsilon^* = \varepsilon' - i\varepsilon'' \tag{2.5}$$

Here, the real part of dielectric permittivity is relative permittivity (ε'). The relative permittivity is the measurement of materials to polarize and store electrical charges when used in capacitors; the imaginary part is dielectric loss (ε') generated during formation and alignment of dipoles. Hourquebie and Olmedo [10] came up with the relationship between dielectric loss and frequency, as in Eq. 2.6:

$$\varepsilon'' \sim f^{-n} \tag{2.6}$$

However, loss factor ($\tan\delta$) is more frequently used for evaluation of dielectric loss, as written in Eq. 2.7:

$$\tan\delta = \varepsilon''/\varepsilon' \tag{2.7}$$

Complex dielectric permittivity is also related to complex electrical conductivity (σ^*). This relationship is given in Eq. 2.8:

$$\sigma^* = \varepsilon''\omega\varepsilon_0 + i\varepsilon'\omega\varepsilon_0 = \sigma' + i\sigma'' \tag{2.8}$$

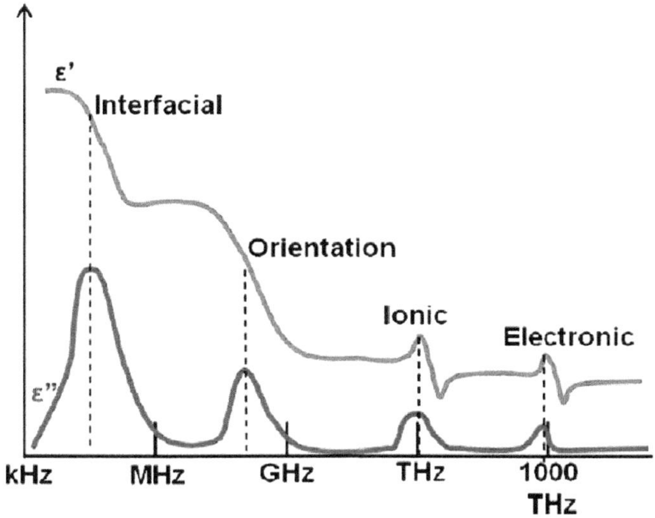

Figure 2.7 Frequency dependent dielectric relaxation of materials showing all four types of polarization mechanisms. Reprinted with permission from Ref. [2], Copyright 2013, American Chemical Society.

Here, ω is the angular frequency, σ' is the real part of complex conductivity (equals to the sum of DC conduction and AC conduction), and σ'' is the imaginary part of AC conductivity. This equation reveals an important relationship between dielectric loss and AC conductivity. High electrical conductivity (more precisely, DC conduction) directly contributes to high dielectric loss. Thus, the strategies that could reduce electrical conduction have been applied to the fabrication of low loss dielectrics [11–16], including multilayer nanocomposites structures with insulating layers and core–shell nanostructures with an insulating shell.

Electric modulus (M^*) is another important parameter for the study of dielectric relaxation of polymer nanocomposites, in particular, the electrically conductive nanocomposites [17, 18]. Electric modulus could be used to suppress the influences of DC conduction on the low frequency dielectric relaxation. The electric modulus is given in Eq. 2.9:

$$M^* = M' + iM'' = \frac{\varepsilon'}{\varepsilon'^2 + \varepsilon''^2} + i\frac{\varepsilon''}{\varepsilon'^2 + \varepsilon''^2} \tag{2.9}$$

Here, M' and M'' are real part and imaginary part of the electric modulus, respectively. M'' is plotted against M' for study of the dielectric relaxation behaviors.

2.3 Dielectric Relaxation Models

2.3.1 Dielectric Relaxation Models for Polymeric Materials

The Debye relaxation model is a classic model for describing the dielectric relaxation of polymer materials, as written in Eq. 2.10:

$$\varepsilon^*(i\omega) = \varepsilon'_\infty + \frac{\Delta\varepsilon'}{1 + i\omega\tau_0} \tag{2.10}$$

Here, $\Delta\varepsilon' = \varepsilon'_s - \varepsilon'_\infty$, ε'_s and ε'_∞ are the relative permittivity at low and high frequency limits and τ_0 is the characteristic relaxation time determined by relaxation frequency. This model neglects the interactions among different dipoles. Currently, dominant models for dielectric relaxation in polymers are variants of the Debye model [19–21].

The Havriliak–Negami model is based on continuous distribution of relaxation time, as given in Eq. 2.11:

$$\varepsilon(i\omega) = \varepsilon'_\infty + \frac{\Delta\varepsilon'}{[1 + (i\omega\tau_0)^{1-\alpha}]^\beta} \tag{2.11}$$

Compared with the Debye model, the two exponents α and β describe the asymmetry and broadness of the relaxation spectra. Typically, the values of both α and β are between 0 and 1.

When $\beta = 1$, the Havriliak–Negami model is reduced to the Cole–Cole model, as given in Eq. 2.12 [22]:

$$\varepsilon(i\omega) = \varepsilon'_{\infty} + \frac{\Delta\varepsilon'}{1+(i\omega\tau_0)^{1-\alpha}} \tag{2.12}$$

When a symmetric dielectric loss peak is obtained from the dielectric relaxation spectra, Eq. 2.12 can be used to describe this relaxation behavior.

Another variation of the Havriliak–Negami model is the so-called Cole–Davidson model, in which, $\alpha = 0$. Thus, this model has a form shown in Eq. 2.13. This model is used when the dielectric loss peak shows asymmetry on dielectric relaxation spectra.

$$\varepsilon(i\omega) = \varepsilon'_{\infty} + \frac{\Delta\varepsilon'}{1+(i\omega\tau_0)^{\beta}} \tag{2.13}$$

The relaxation parameters in these models can be extracted upon fitting the experimental data, providing more profound understanding of dielectric relaxation behaviors for polymer materials. However, it needs to point out that understanding of physical meanings of these parameters is very limited, although there has been already great success achieved for various polymer systems. For example, the Havriliak–Negami model is only an empirical modification of the classic Debye relaxation model.

2.3.2 Models for Polymer Nanocomposites

Interfacial polarization is the main characteristic in dielectric relaxation of polymer nanocomposites. In the field of polymer nanocomposites, this polarization mechanism is also called the Maxwell–Wagner–Sillar polarization. In this mechanism, a double-layer model was proposed for describing the interfacial behaviors in the heterogeneous system, such as polymer nanocomposites. The simplest form of mathematical expression of this mechanism is given in Eq. 2.14, which is also called the Maxwell–Wagner equation [21]:

$$\widetilde{\varepsilon^{*}}(i\omega) = \tilde{\varepsilon}'_{\infty} + \frac{\Delta\tilde{\varepsilon}'}{1+i\omega\tau_{\mathrm{MW}}} \tag{2.14}$$

$$\tau_{\mathrm{MW}} = \varepsilon_0 \frac{\varepsilon'_1 + \varepsilon'_2}{\sigma'_1 + \sigma'_2} \tag{2.15}$$

When both dielectric layers have the same spacing in this double layer model, we have

$$\tilde{\varepsilon}'_\infty = \frac{\varepsilon'_1 \varepsilon'_2}{\varepsilon'_1 + \varepsilon'_2} \tag{2.16}$$

$$\Delta\tilde{\varepsilon}' = \frac{\varepsilon'_1 \sigma'_2 + \varepsilon'_2 \sigma'_1}{(\sigma'_1 + \sigma'_2)^2 (\varepsilon'_1 + \varepsilon'_2)} \tag{2.17}$$

Here, 1 and 2 represent the polymer and nanomaterial, respectively, when this model is used for polymer nanocomposites; σ' and ε' are the real part of complex conductivity and dielectric permittivity, respectively; and τ_{MW} is the relaxation time of this double-layer structure.

However, there is a limitation in this mechanism. First of all, this mechanism fits better in polymer nanocomposites with low volume fractures. Secondly, it was derived from spherical inclusions, thus, its accuracy on analyzing nanomaterials, such as nanofibers and nanotubes, may be unsatisfactory.

Polymer nanocomposites are multiphase material systems, thus, the compositions, properties, and morphologies of the dispersed phase, i.e., nanomaterials, have significant impacts on the performances of the resulting nanocomposites. Besides, efforts on the aforementioned relaxation models, theoretical analysis of dielectric properties have been attempted to establish correlations between the properties of polymers, nanomaterials and nanocomposites via their volume fractions. Thus, effective medium theories and their derivations are frequently applied for this purpose. However, similar to the Maxwell–Wagner–Sillar polarization mechanisms, the effective medium theories and their derivations, in the case of dielectric properties, are limited by their insufficiency in analyzing nanocomposites with (1) high volume fraction of nanomaterials and/or (2) non-spherical/non-ellipsoidal nanomaterials. Moreover, for the polymer nanocomposites modified by highly conductive nanomaterials, such as metals and carbon nanomaterials, the application of effective medium theories has limitation. First of all, the metals usually have negative dielectric permittivity in the low frequency range; secondly, understanding of dielectric properties of carbon nanomaterials is still insufficient. Thus, the efficiency of the effective medium theories is questionable in nanocomposites.

The Maxwell–Garnett model is one of the most simple and convenient derivations of the effective medium theories. It can provide nice accuracy at low loading levels of nanomaterials in polymer nanocomposites. The mathematical equation for binary system is given in Eq. 2.18:

$$\frac{\varepsilon_m - \varepsilon_n}{2\varepsilon_m + \varepsilon_n} = V_f \frac{\varepsilon_m - \varepsilon_f}{2\varepsilon_m + \varepsilon_f} \tag{2.18}$$

The Bruggeman model is another convenient form of the effective medium theory and very useful for describing the systems with aggregate topology, and the equation of modeling binary system is given in Eq. 2.19:

$$V_m \frac{\varepsilon_m - \varepsilon_n}{2\varepsilon_m + \varepsilon_n} + V_f \frac{\varepsilon_f - \varepsilon_n}{2\varepsilon_f + \varepsilon_n} = 0 \tag{2.19}$$

Where n, m, and f represent nanocomposites, matrix, and fillers, respectively, and V is the volume fraction.

A general form of the effective medium theory is given in Eq. 2.20, which could be used for hybrid polymer nanocomposites containing more than one type of nanomaterials or polymers in the nanocomposites, as follows:

$$\frac{\varepsilon_m - \varepsilon_n}{2\varepsilon_m + \varepsilon_n} = \sum_i V_{fi} \frac{\varepsilon_m - \varepsilon_{fi}}{2\varepsilon_m + \varepsilon_{fi}} \tag{2.20}$$

As mentioned earlier, these theories have provided reasonably good accuracy when predicting the dielectric properties of the nanocomposites. However, they constantly failed at high volume fraction or in the nanocomposites containing nanomaterials with high aspect ratio, because these theories do not consider the particle–particle dipolar interactions as well as the morphologies of nanomaterials [23]. Also, the addition of rigid nanomaterials often leads to voids. Therefore, many more modified effective medium theories have been proposed to obtained high prediction accuracy, such as self-consistent effective medium theory [24], multi-core model [25], theoretical interlayer model [26], and double-inclusion model [27].

Jaysundere–Smith equation, as shown in Eqs. 2.21–2.23 [23, 28, 29], is a frequently used modified effective medium theory. In this modified equation, the dipolar interactions between adjacent spherical particles were considered. In this model, each spherical particle was treated as a polarizable particle, within which the reduction of electric field was considered.

$$\varepsilon_n = \frac{\varepsilon_m V_m + \varepsilon_f V_f (A)(B)}{V_m + V_f (A)(B)} \tag{2.21}$$

$$A = \frac{3\varepsilon_m}{2\varepsilon_m + \varepsilon_f} \tag{2.22}$$

$$B = 1 + \frac{3V_f(\varepsilon_f - \varepsilon_m)}{2\varepsilon_m + \varepsilon_f} \tag{2.23}$$

In Yamada's model [28], as given in Eq. 2.24, a morphology factor (p) dependent on both shapes and orientation of ellipsoidal particles was introduced. The value of this morphology factor should be between 0 and 1.

$$\varepsilon_n = \varepsilon_m \left(1 + \frac{V_f(\varepsilon_f - \varepsilon_m)}{\varepsilon_p + p(1 - V_f)(\varepsilon_f - \varepsilon_m)} \right) \tag{2.24}$$

2.4 High-Field Dielectric Relaxation

Dielectric relaxation spectroscopy has shown great success in measurement of dielectric relaxation behaviors of materials in a low electrical field. During the characterization process, an applied voltage is usually around 1–5 V. However, the same technique is not ideal for describing the dielectric relaxation behaviors at higher electric fields up to several hundreds of kilovolts per centimeter because of the instrument's limitation in analyzing dipole switching at higher fields. Instead of the dielectric relaxation spectroscopy, a dielectric displacement (D)–electric field (E) relationship has been applied to characterize the dielectric relaxation behaviors in the presence of high electric field. The dielectric properties are significantly affected by the magnitude of applied electric field, in particular, high electric fields. Meanwhile, understanding of dielectric relaxation behaviors under high electric fields is also critical to high power applications, such as high voltage pulse capacitors.

In a parallel plate capacitor, as shown in Fig. 2.6, the dielectric displacement (D) is defined as the surface charge density, i.e., the amount of charges stored per unit area on the plate. In the presence of a dielectric medium, the dielectric displacement is directly proportional to the magnitude of electric field (E), described by Eq. 2.25:

$$D = \varepsilon_0 E + P = \varepsilon_0 \varepsilon_r E \quad (2.25)$$

Where the polarization P is determined by Eq. 2.2:

$$P = \varepsilon_0(\varepsilon_r - 1)E \quad (2.26)$$

The relationships between dielectric displacement (or polarization) and electric field can be recorded by a ferroelectric tester, and a typical monopolar D–E hysteresis loop is given in Fig. 2.8. According to Fig. 2.8, the dielectric properties at various electric fields can be obtained. More importantly, the D–E loop is an excellent measurement of energy storage density (U), according to Eq. 2.13:

$$U = \int E dD \quad (2.27)$$

Mathematically, U = released energy + unrecovered energy. While the released energy is desired, the unrecovered energy is considered as a form of energy loss. For a nonpolar polymer material without permanent dipoles, such as polyethylene (PE) and polypropylene (PP), its unrecovered energy is nearly zero and the D–E hysteresis loop shows a nearly linear relationship, in opposite to PVDF and some other polar materials, suggesting a low energy loss. For this type of linear dielectric materials, Eq. 2.27 can be modified to Eq. 2.28:

$$U = \frac{1}{2}\varepsilon_0 \varepsilon_r E^2 \quad (2.28)$$

However, compared with polymer materials with permanent dipoles and large energy loss, such as PVDF, the dielectric displacement of PE and PP at the same electric field is much smaller, implying less capability of storing energy. Therefore, the balance between high energy storage and low energy loss is an important issue to be addressed. Normal ferroelectric polymers, such as PVDF, usually possess larger energy loss, although they also have high energy storage. To achieve both high energy storage and low energy loss, materials showing relaxor ferroelectric and antiferroelectric behaviors are critical. As shown in Fig. 2.9, both hysteresis behaviors lead to high energy storage and high energy efficiency. Therefore, relaxor ferroelectric and antiferroelectric polymers are highly desired for polymeric materials with both high energy storage capability and high energy efficiency. The relaxor ferroelectric behavior has been found in terpolymers of PVDF, which will be discussed in Chapter 6.

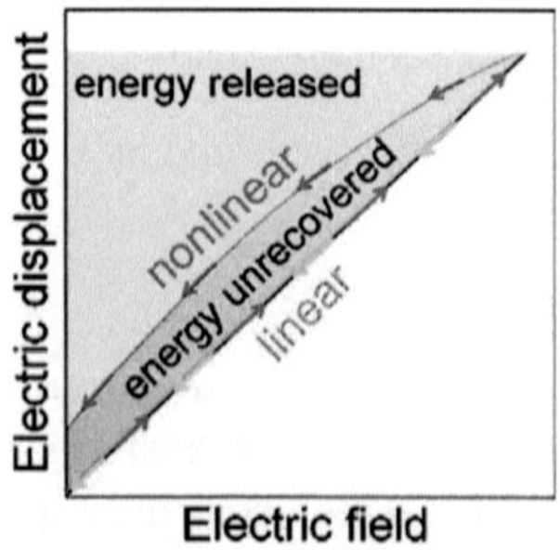

Figure 2.8 Schematic representation of unipolar electric displacement–electric field (D–E) hysteresis loops for both linear and nonlinear dielectric polymers under high-field switching. Reprinted with permission from Ref. [1], Copyright 2012, American Chemical Society.

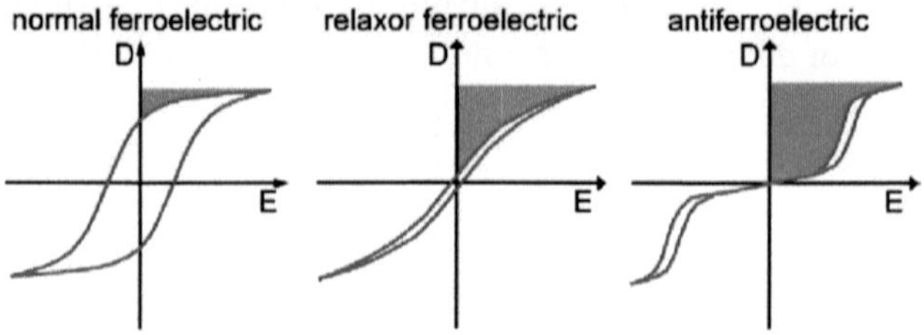

Figure 2.9 Comparison of various nonlinear electric displacement–electric field (D–E) hysteresis behaviors. Reprinted with permission from Ref. [1], Copyright 2012, American Chemical Society.

References

1. Zhu, L., and Q. Wang (2012). Novel ferroelectric polymers for high energy density and low loss dielectrics. *Macromolecules*, **45**(7):2937–2954.
2. Fredin, L. A., et al. (2013). Sustainable high capacitance at high frequencies: Metallic aluminum–polypropylene nanocomposites. *ACS Nano*, **7**(1):396–407.

3. Frubing, P., F. P. Wang, and M. Wagener (2012). Relaxation processes and structural transitions in stretched films of poly(vinylidene fluoride) and its copolymer with hexafluoropropylene. *Appl Phys A: Mater Sci Process*, **107**(3):603–611.
4. El Sayed, A. M., and W. M. Morsi (2013). Dielectric relaxation and optical properties of polyvinyl chloride/lead monoxide nanocomposites. *Polym Compos*, **34**(12):2031–2039.
5. Shukla, N., et al. (2014). Dielectric relaxation and thermal studies on dispersed phase polymer nanocomposite films. *J Mater Sci: Mater Electron*, **25**(6):2759–2770.
6. Zheng, S. Y., et al. (2014). Oxygen vacancy-related dielectric relaxation and electrical conductivity in La-doped Ba(Zr0.9Ti0.1)O-3 ceramics. *J Mater Sci: Mater Electron*, **25**(9):4058–4065.
7. Abdelkader, K., et al. (2015). Dielectric relaxation and ionic conduction in 66%Silica/CW229-3/HW229-1 microcomposite polymer. *Composites Part B*, **78**:488–496.
8. Belovickis, J., et al. (2015). Dielectric spectroscopy of polymer based PDMS nanocomposites with ZnO nanoparticles. *Ferroelectrics*, **479**(1):82–89.
9. Svirskas, S., et al. (2015). Dielectric relaxation and ferromagnetic resonance in magnetoelectric (polyvinylidene-fluoride)/ferrite composites. *J Polym Res*, **22**(7).
10. El Ghanem, H. M., et al. (2013). Effect of dc-bias on the dielectric behavior of CNT/ABS nanocomposites. *Phys B: Condens Matter*, **418**:41–46.
11. Sun, L. L., et al. (2010). Structure-induced high dielectric constant and low loss of CNF/PVDF composites with heterogeneous CNF distribution. *Nanotechnology*, **21**(30).
12. Yang, K., et al. (2012). Core-shell structured polystyrene/$BaTiO_3$ hybrid nanodielectrics prepared by in situ RAFT polymerization: A route to high dielectric constant and low loss materials with weak frequency dependence. *Macromol Rapid Commun*, **33**(22):1921–1926.
13. Xie, L. Y., et al. (2011). Core-shell structured poly(methyl methacrylate)/$BaTiO_3$ nanocomposites prepared by in situ atom transfer radical polymerization: A route to high dielectric constant materials with the inherent low loss of the base polymer. *J Mater Chem*, **21**(16):5897–5906.
14. Xie, L. Y., X. Y. Huang, and P. K. Jiang (2011). Low loss and high dielectric constant poly(methyl methacrylate)/$BaTiO_3$ nanocomposites

prepared by in situ atom transfer radical polymerization. Proceedings of 2011 International Conference on Electrical Insulating Materials (Iseim, 2011), pp. 92–94.

15. Tang, Z. C. W., et al. (2012). Structurally induced dielectric constant promotion and loss suppression for poly(vinylidene fluoride) nanocomposites. *Macromol Mater Eng*, **297**(5):420–426.
16. Huang, X. Y., et al. (2014). Core-double-shell structured nanocomposite dielectrics with high permittivity and low loss for electric energy storage. Proceedings of 2014 International Symposium on Electrical Insulating Materials (Iseim, 2014), pp. 22–25.
17. Yang, W. H., et al. (2011). Electrical modulus analysis on the Ni/CCTO/PVDF system near the percolation threshold. *J Phys D: Appl Phys*, **44**(47).
18. Saji, J., et al. (2015. Impedance analysis, dielectric relaxation, and electrical conductivity of multi-walled carbon nanotube-reinforced silicon elastomer nanocomposites. *J Elastomers Plast*, **47**(5):394–415.
19. Williams, G. (2005). Polymer dynamics and broadband dielectric spectroscopy, in *New Polymeric Materials*, ed. L. S. KorugicKarasz, W. J. MacKnight, and E. Martuscelli. Washington: American Chemical Society, pp. 268–281.
20. Asami, K. (2002). Characterization of heterogeneous systems by dielectric spectroscopy. *Prog Polym Sci*, **27**(8):1617–1659.
21. Petzelt, J., et al. (2013). Broadband dielectric and conductivity spectroscopy of inhomogeneous and composite conductors. *Phys Status Solidi A*, **210**(11):2259–2271.
22. Wu, Y., et al. (2015). Exceptional dielectric properties of chlorine-doped graphene oxide/poly (vinylidene fluoride) nanocomposites. *Carbon*, **89**:102–112.
23. Kim, P., et al. (2009). High energy density nanocomposites based on surface-modified $BaTiO_3$ and a ferroelectric polymer. *ACS Nano*, **3**(9):2581–2592.
24. Li, J. Y., L. Zhang, and S. Ducharme (2007). Electric energy density of dielectric nanocomposites. *Appl Phys Lett*, **90**(13):132901.
25. Tanaka, T., et al. (2005). Proposal of a multi-core model for polymer nanocomposite dielectrics. *IEEE Trans Dielectr Electr Insul*, **12**(4):669–681.
26. Steeman, P. A. M., and F. H. J. Maurer (1992). An interlayer model for the complex dielectric constant of composites: An extension to ellipsoidally shaped particles. *Colloid Polym Sci*, **270**(11):1069–1079.

27. Hori, M., and S. Nemat-Nasser (1993). Double-inclusion model and overall moduli of multi-phase composites. *Mech Mater*, **14**(3):189–206.

28. Choudhury, A. (2012). Preparation, characterization and dielectric properties of polyetherimide nanocomposites containing surface-functionalized $BaTiO_3$ nanoparticles. *Polym Int*, **61**(5):696–702.

29. Yang, K., et al. (2013). Fluoro-polymer@$BaTiO_3$ hybrid nanoparticles prepared via raft polymerization: Toward ferroelectric polymer nanocomposites with high dielectric constant and low dielectric loss for energy storage application. *Chem Mater*, **25**(11):2327–2338.

Chapter 3

Perovskite Ceramics and Their Polymer Nanocomposites

Bin Li[a] **and Wei-Hong Zhong**[b]
[a]*Department of Mechanical Engineering, Wichita State University, 1845 Fairmount Street, Wichita, Kansas 67260-0133, USA*
[b]*School of Mechanical and Materials Engineering, Washington State University, 405 NE Spokane Street, Pullman, Washington 99164-2920, USA*
bin.li@wichita.edu, katie_zhong@wsu.edu

Many inorganic ceramics have large permanent dipole moment, due to their chemical compositions and crystal structures. They typically exhibit very high dielectric permittivity up to a few thousands, in contrast to the low dielectric permittivity of polymer materials. However, with increasing demand of flexible, embedded, and miniaturized electronic devices, these ceramic materials alone fail to achieve similar success as in traditional electronics. Some reasons behind this include high brittleness, high mass density, low dielectric breakdown strength as well as poor processability of ceramic materials, and so on, compared with polymers. Thus, incorporating dielectric ceramics into polymers to fabricate polymer composites/

Polymer Nanocomposites for Dielectrics
Edited by Wei-Hong Zhong and Bin Li

ISBN 978-981-4745-53-6 (Hardcover), 978-981-4745-54-3 (eBook)
www.panstanford.com

nanocomposites has been adopted to combine the advantages of both types of materials [13, 34, 94].

Metal oxides with perovskite crystal structures have been drawing tremendous attentions from both academia and industry for dielectric polymer nanocomposite applications, because of their high dielectric permittivity, ferroelectricity, piezoelectricity, pyroelectricity, and so on. Popular perovskite type of metal oxides for dielectric polymer nanocomposites include barium titanate ($BaTiO_3$), lead zirconate titanate, $Pb[Zr_xTi_{1-x}]O_3$ (PZT, $0 < x < 1$), and their doped ceramic structures [26, 34, 35, 41, 52, 53, 60, 61]. These perovskite-type metal oxides show multiple crystal phases altered by temperature and particle size, leading to different dielectric properties. In this chapter, perovskite ceramic materials will be introduced, with a special focus on $BaTiO_3$. In particular, besides the discussion on the factors such as concentrations, particle size, and the morphologies of nanoparticles, special attention will be paid to the interfacial structures, namely, the surface modification of the ceramic nanomaterials, and their effects on dielectric performances of the polymer nanocomposites.

3.1 Crystal Structures of Barium Titanate

Barium titanate ($BaTiO_3$) is one of the most important perovskite-type ceramic materials with various crystal structures. The strong permanent dipole moments in some $BaTiO_3$ crystals result in high dielectric permittivity and ferroelectric behaviors. Meanwhile, $BaTiO_3$ exhibits remarkable piezoelectricity and pyroelectricity as well. The outstanding dielectric performances of $BaTiO_3$ have popularized it in polymer composite dielectric films.

The dielectric performances of $BaTiO_3$ are directly related to its crystal structures. $BaTiO_3$ is a type of perovskite oxide with a chemical formula of ABO_3 (A = Ba, Sr . . .; B = Ti, Zr . . .), the crystal structure of which consists of AO_{12} cuboctahedra and BO_6 octahedra (Fig. 3.1). $BaTiO_3$ shows five crystal phases, and they are rhombohedral, orthorhombic, tetragonal, cubic, and hexagonal perovskite crystal structures [19, 36, 52, 61]. The phase transition among different crystal structures is a function of both temperature and particle size. Typically, the rhombohedral to orthorhombic transition occurs at 183 K, the orthorhombic phase to tetragonal transition occurs at 278 K, and the tetragonal to cubic transition occurs at 403 K [36, 68]. At a high temperature of 1705 K, the cubic-to-hexagonal phase transition will happen [36].

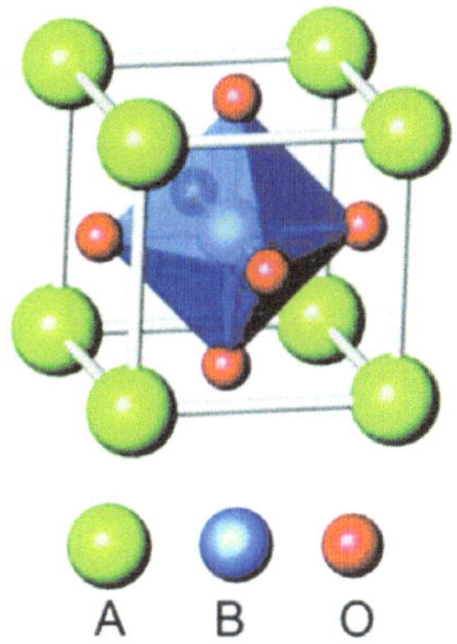

Figure 3.1 Schematic sketches of perovskite structure of ABO_3-type ceramics. In case of $BaTiO_3$, A = Ba; B = Ti. Reproduced from Ref. [61] with permission of The Royal Society of Chemistry.

In addition to temperature, the particle size has proven to be another essential factor to crystal structures [3, 12, 19, 27, 43, 68]. The changes in particle size can affect the stability of crystal phases and result in phase transition. It has been reported that with decrease in particle size, the tetragonal-to-cubic transition can also take place at temperature lower than 403 K, such as room temperature [3, 12, 19, 36, 43]. However, the critical particle size in this size-dependent phase transition usually vary depending on different synthesis techniques, such as hydrothermal method, sol–gel method, and decomposition method [6, 36, 52, 69]. The synthesis methods do not only affect crystal structures of $BaTiO_3$, but also cause differences in their internal stresses, structural defects, and impurities. These factors can contribute to the phase transition, too [71].

Among these five phases, cubic crystal phase is an ideal centrosymmetric perovskite structure. Thus, there is no net dipole moment in the cubic crystal phase, showing paraelectricity (Fig. 3.2). The dipole moment only exists when electric field is applied. While all other four $BaTiO_3$ crystal phases exhibit ferroelectricity due to permanent dipole moment from their noncentrosymmetric structures. For example, in the tetragonal structure (Fig. 3.2), the permanent dipole moment exists along *c*-axis direction, because of the relative displacement of Ti. Meanwhile, owing to the room temperature ferroelectricity and maximum permittivity of tetragonal $BaTiO_3$, it has been of great interests in development of $BaTiO_3$ for both dielectric ceramics and dielectric composites [61]. These phase-dependent dielectric properties also result in the size dependent dielectric properties of $BaTiO_3$ [72–74, 85, 92]. Typically, with the reduction of particle size of $BaTiO_3$, remarkable decrease in dielectric permittivity can be observed [19]. The phase transition to cubic perovskite structure with lower dielectric permittivity is considered as one major reason for this reduction. Meanwhile, the size-dependent "dilution" effect of a grain boundary "dead" layer was proposed as another reason for this reduction in dielectric permittivity [92]. But it should be pointed out here, although the tetragonal phase possesses attractive dielectric permittivity, its ferroelectric behavior also suggests large dielectric loss. In contrast, the cubic crystal phase behaves similarly to a linear dielectric with low dielectric loss. Thus, the paraelectric cubic $BaTiO_3$ is still of great importance in many applications.

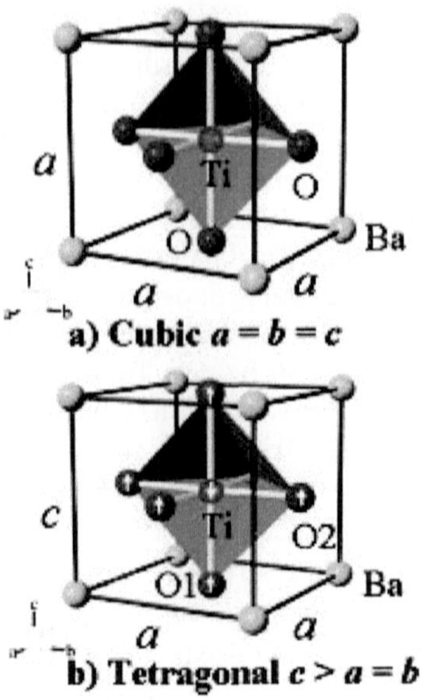

Figure 3.2 Schematic sketches of (a) cubic perovskite crystal structure and (b) tetragonal perovskite crystal structure in $BaTiO_3$. Reprinted with permission from Ref. [36], Copyright 2006, American Chemical Society.

3.2 Barium Titanate /Polymer Nanocomposites

With the increase of $BaTiO_3$ loading in polymer matrix, the increase in dielectric permittivity and dielectric loss are expected [20, 21]. Effective medium theories as well as the Maxwell–Wagner (MW) equations introduced in Chapter 2 have achieved fairly good success in evaluation of dielectric relaxation of dielectric ceramic/polymer nanocomposites, suggesting strong composition dependence of dielectric properties of this class of polymer nanocomposites. This is also true for polymer nanocomposites reinforced by other types of dielectric ceramics.

In order to achieve high permittivity, the concentration of $BaTiO_3$ often went up to 50 wt% or even higher [48, 54, 76, 94]. However, the large deviation typically occurs at high concentration levels between measured and calculated dielectric permittivity. One example is given in Fig. 3.3. In this study, the Lichtenecker's logarithmic mixing rule was applied to predict the dielectric permittivity of the mixture, as shown in Eq. 3.1.

$$\ln \varepsilon_c = \phi_f \ln \varepsilon_f + \phi_m \ln \varepsilon_m \tag{3.1}$$

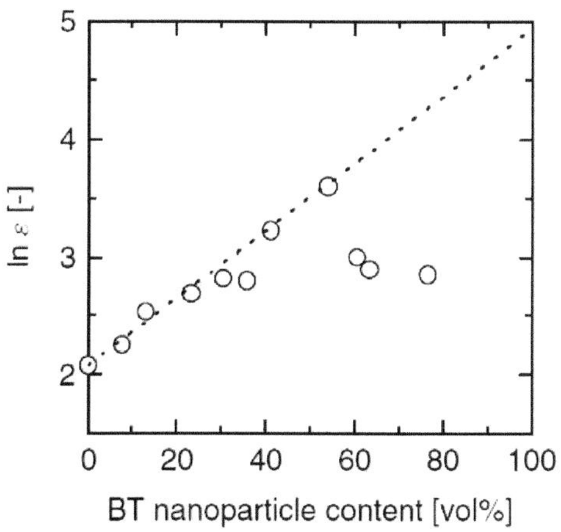

Figure 3.3 Semi-logarithmic plot of dielectric constant at various $BaTiO_3$ (BT nanoparticle contents). The dotted line shows the calculated values from Lichtenecker's logarithmic mixture formula. Reproduced from Ref. [54] with permission from John Wiley and Sons.

Regarding the high concentration level (e.g., higher than 50%), besides the limited accuracy of these models, the aggregation phenomena of nanoparticles is considered as a key factor to this deviation. Meanwhile, the addition of a large amount of ceramic particles in the polymer matrix incurs more voids to be formed, which can reduce dielectric permittivity and dielectric breakdown strength of the composites. The similar deviation was also found in other polymer/$BaTiO_3$ nanocomposites studied from different models [45, 76].

As discussed earlier, the dielectric performances of $BaTiO_3$ particles are size dependent. In general, the dielectric permittivity of $BaTiO_3$ nanoparticles decreases as the particle size reduces, due to the phase transition from tetragonal to cubic phase. A similar size dependence of dielectric properties was found in poly(vinylidene fluoride) (PVDF)/$BaTiO_3$ nanocomposites [25, 26]. The study reported a remarkable dependence of the dielectric properties of the nanocomposites on particle size, as shown in Fig. 3.4. With the increase in particle size, both dielectric permittivity and dielectric loss of the nanocomposites increased. However, it is necessary to argue that this particle size dependent dielectric properties of the nanocomposites might be related to the crystal structures of PVDF and the concentration of $BaTiO_3$ in the nanocomposites, on which the authors did not provide sufficient discussion. In particular, PVDF is well known for the polymorphs of multiple crystal structures, as well as their significant impacts on dielectric polarizations, thus, the understanding of the effects of particle size on the crystal structures of PVDF seems to be necessary. In another study conducted by the same research group on polyimide (PI)/$BaTiO_3$ nanocomposites, some interesting results were reported, as shown in Fig. 3.5. Obviously, the particle size dependence of the dielectric permittivity of PI/$BaTiO_3$ nanocomposites is affected by the volume fraction of $BaTiO_3$ as well. However, large particles with very high dielectric permittivity did not further lead to the highest dielectric permittivity of resulting polymer nanocomposites. This result was not completely unexpected, since so many factors contribute to dielectric properties of polymer nanocomposites. This complexity is also responsible for the inconsistence among reported size dependent dielectric properties of PVDF/$BaTiO_3$ nanocomposites [18, 31, 49, 50].

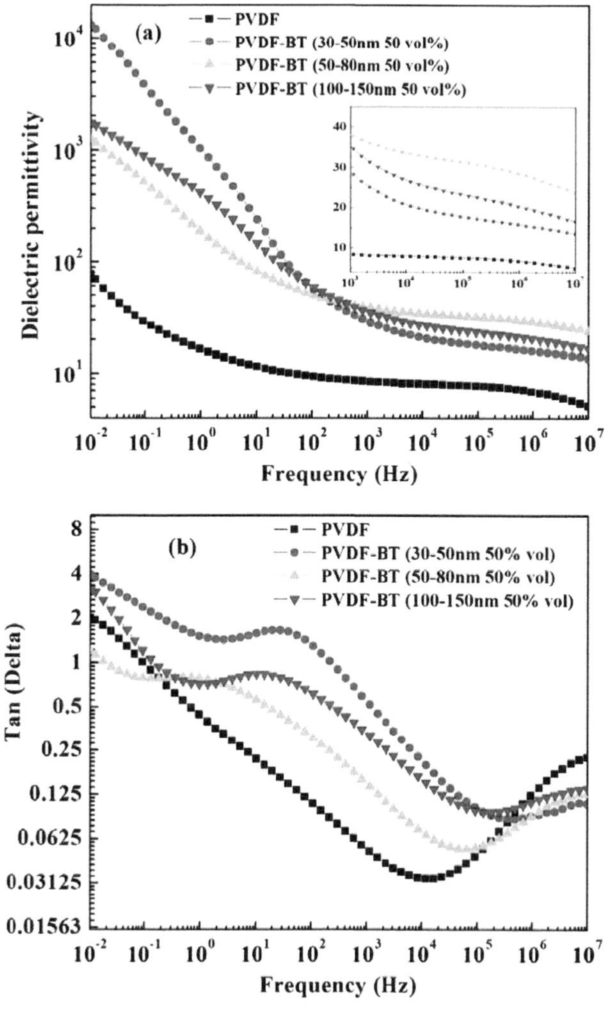

Figure 3.4 Dielectric properties of PVDF/$BaTiO_3$ nanocomposites with different $BaTiO_3$ particle sizes. Reprinted with permission from Ref. [25], Copyright 2012, AIP Publishing LLC.

In addition to the dielectric properties and concentrations of ceramic nanomaterials, the dispersion as well as the interface (i.e., surface modification of ceramic nanomaterials) are also critical to the dielectric relaxation behaviors [1, 7, 9, 16, 34, 40, 46, 56]. In particular, with the decrease in particle size down to nanoscale, the surface area of ceramic particles rapidly increases, leading to very high surface energy and tendency to aggregate, as well as high interfacial areas in polymer nanocomposites. Both factors can cause big impacts on the interfacial polarization in nanocomposites. This is probably the reason for the increased deviation between theoretical properties from various models and the experimental values (Fig. 3.5) with the volume fraction of nanoparticles. Furthermore, in order to achieve polymer nanocomposites with high dielectric performances, different $BaTiO_3$ nanostructures and hybrid nanomaterial systems have also been applied.

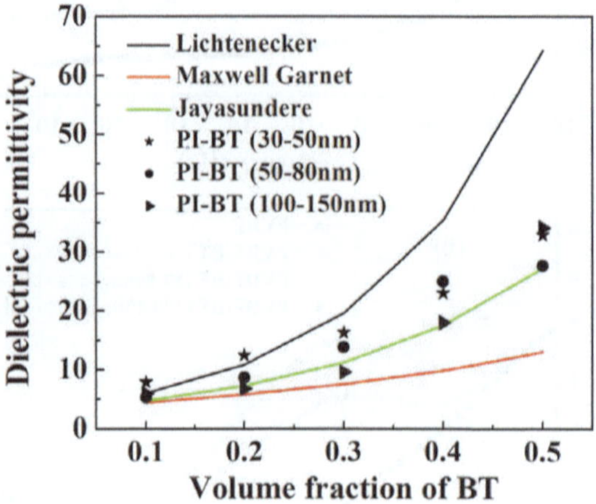

Figure 3.5 Theoretical and experimental dielectric properties of polyimide/ $BaTiO_3$ nanocomposites with different $BaTiO_3$ particle sizes. Reprinted with permission from Ref. [24], Copyright 2012, AIP Publishing LLC.

3.3 Surface Modification of $BaTiO_3$ in Polymer Nanocomposites

Surface modification has always been a crucial issue in fabrication of high-performance polymer nanocomposites. It not only matters to the dispersion of nanomaterials and interfacial bonding between polymer matrix and nanomaterials, but also plays an important role in dielectric relaxation in nanocomposites by impacting interfacial polarization [66]. Furthermore, the surface-modified nanoparticles often show altered dielectric properties of pristine nanoparticles. In a comparative study of dielectric properties of unmodified and modified $BaTiO_3$ nanoparticles, *n*-hexylphosphonic acid (HPA) was used to modify $BaTiO_3$ nanoparticles (6 nm in size) [13]. As shown in Fig. 3.6, the surface modification led to significant changes in the dielectric properties: the modified nanoparticles have lower dielectric constant and dielectric loss, as well as weaker temperature sensitivity.

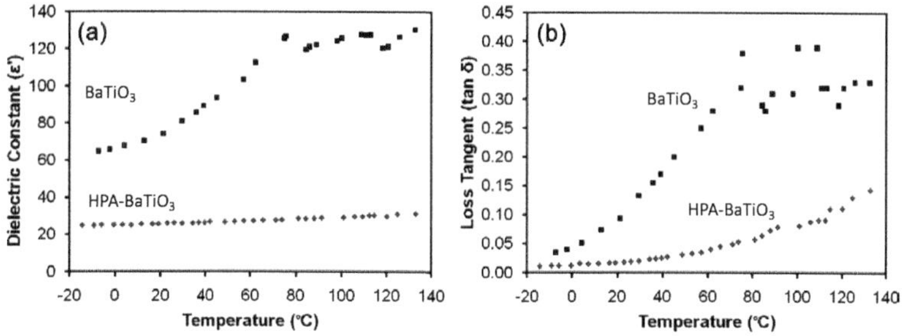

Figure 3.6 Schematic sketches of the effect of 1D nanostructure on the electrical conductivity of polymer nanocomposite. Reprinted with permission from Ref. [13], Copyright 2010, American Chemical Society.

3.3.1 Surface Modification via Organic Molecules

$BaTiO_3$ particles have polar and reactive surfaces because of surface groups such as hydroxyl. This provides convenience for grafting different types of surfactants onto the particle surface and form a dielectric layer. In the meantime, to increase the grafting rate of surfactants, the $BaTiO_3$ particles are typically hydroxylated in H_2O_2 at elevated temperatures to increase the reactive sites on the particle surface [4, 8, 88], as shown in Fig. 3.7. The resulting nanocomposites show lowered dielectric permittivity and dielectric loss, compared with the nanocomposites modified by pristine nanoparticles; however, the surface hydroxylation successfully improved dielectric strengths. The authors believed the changes in dielectric properties were owing to the strong interactions between hydroxyl groups and polymer matrix (PVDF). A similar type of hydroxylated $BaTiO_3$ nanoparticles was also applied to another polymer matrix, polyetherimide (PEI) [14]. In this study, the surface hydroxylation made contributions to the dielectric permittivity. The hydroxylated $BaTiO_3$ nanoparticle led to higher dielectric permittivity of PEI nanocomposites, compared with pristine $BaTiO_3$

nanoparticles. Both pristine $BaTiO_3$ and hydroxylated $BaTiO_3$ could be modified by surfactants, such as various silane coupling agents [30, 32, 33, 38, 47, 89]. The popularity of silane coupling agents in modifying nanomaterials is from its high reactivity and diverse functional groups to interact or react with the polymer matrix. Figure 3.8 shows some examples of the silane coupling agents for surface modification of nanomaterials. The functional groups can remarkably change the surface properties of $BaTiO_3$ nanoparticles and affect the dispersion of nanoparticles as well as the interfacial properties in the nanocomposites. The studies on effects of different silane coupling agents on the performances of $BaTiO_3$/epoxy nanocomposites [32, 33] showed that the silane coupling agents played a critical role in their thermal and dielectric properties. By using proper silane coupling agents, the desired dielectric performances of the nanocomposites could be achieved. For example, both BT-OH and BT-HBP could enhance the dielectric permittivity of the nanocomposites, while BT-EP, BT-SH, and BT-NH could lower the dielectric permittivity to different levels. Similar effects were found in high electric field dielectric properties as well, suggesting the effects of surface properties of nanoparticles on dielectric properties of polymer nanocomposites.

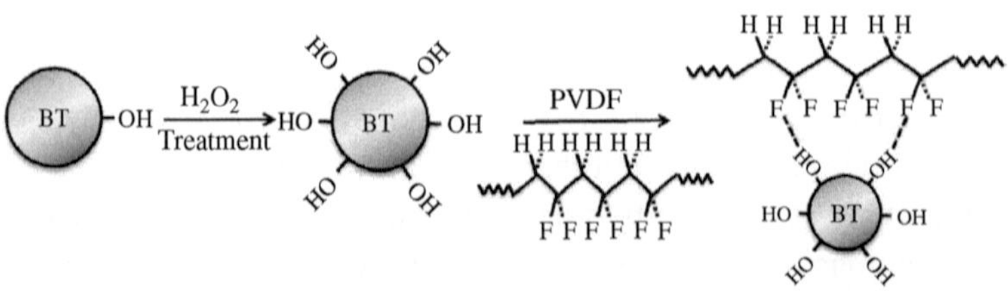

Figure 3.7 Schematic sketches of hydroxylation of $BaTiO_3$ particles. Reprinted with permission from Ref. [93], Copyright 2011, American Chemical Society.

In addition, there are some other coupling agents applied to the modification for nanoparticles [39]. For example, in Fig. 3.9, the tinanate coupling agent (DN 101) was used to modify both pristine and hydroxylated $BaTiO_3$ nanoparticles [28]. This study suggested that hydroxyl groups on the particle surface could lead to an enhanced charge accumulation, which was good for increasing polarization and energy density. Moreover, the synergetic effects between coupling agents and hydroxyl groups improved the interfacial polarization in the resulting polymer nanocomposites, leading to better dielectric performances.

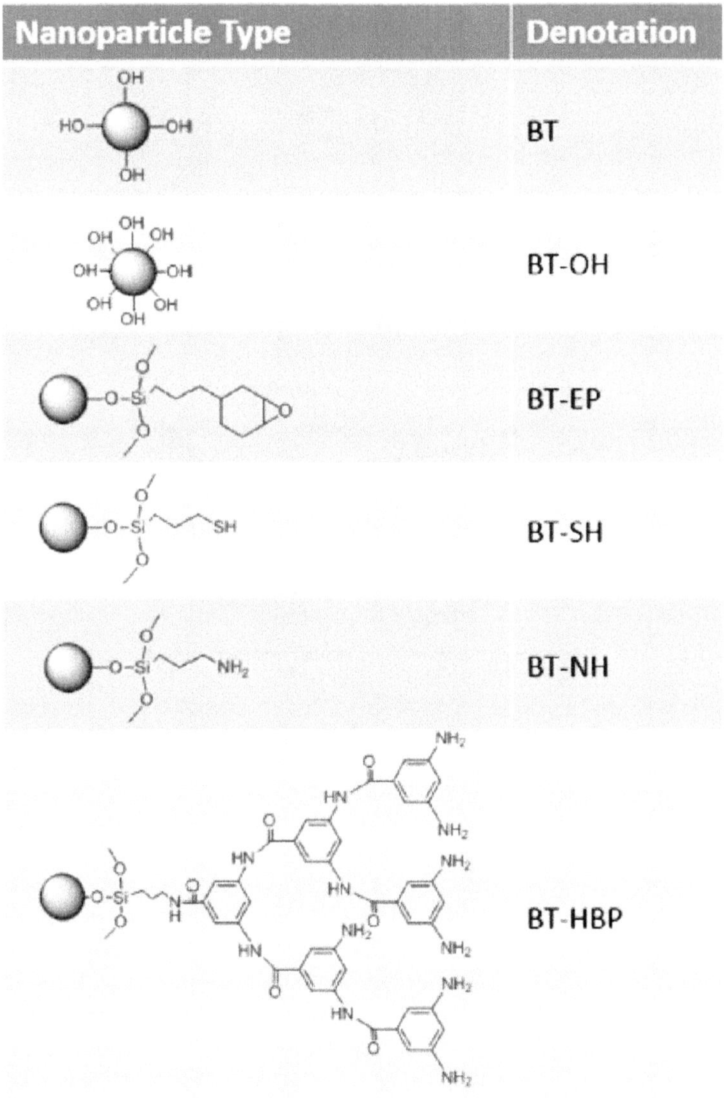

Figure 3.8 Schemes of as-received and silane modified $BaTiO_3$ nanoparticles. Reprinted with permission from Ref. [33], Copyright 2014, IEEE.

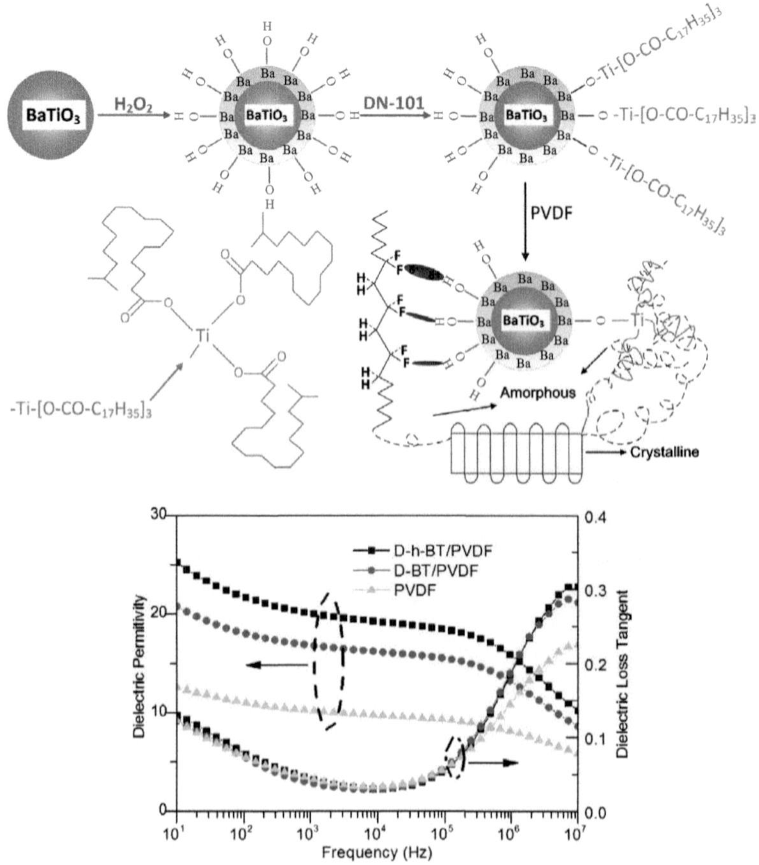

Figure 3.9 Surface modification of $BaTiO_3$ nanoparticle via titanate coupling agent (DN-101) for PVDF/$BaTiO_3$ nanocomposites and its effects on dielectric properties. D-BT: DN-101-modified pristine $BaTiO_3$; D-h-BT: DN-101-modified hydroxylated $BaTiO_3$. Reprinted with permission from Ref. [28], Copyright 2014, American Chemical Society.

Phosphonates and phosphonic acids are another type of effective surfactants to modify $BaTiO_3$ [22, 84]. They have strong binding affinity to oxides [13]. At the same time, they do not have the issues that occur to sliane coupling agents, that is, self-condensation of silane molecules as well as reactions with water. Figure 3.10 presents surface modification of cubic $BaTiO_3$ nanoparticles using a fluorinated phosphonic acid ligand: pentafluorobenzyl phosphonic acid (PFBPA) [40]. According to the study, the phosphonic acid formed well-ordered and tightly bound monolayers. With increasing loading of surface-modified $BaTiO_3$ nanoparticles, the increase in dielectric permittivity was observed until the loading exceeded 50 vol%, which was believed to be resulted from the increased porosity of the nanocomposites. Compared with theoretical dielectric properties calculated by using of different models, the higher accuracy was achieved at low loading levels. These models failed in describing dielectric properties of the nanocomposites with high loadings of nanoparticles, because of such factors as particle aggregation, porosity of polymer nanocomposites, and so on. However, in this study, the authors did not provide the dielectric permittivity of the untreated $BaTiO_3$-reinforced polymer nanocomposites. Thus, the contribution of the surfactant (PFBPA) remains unknown.

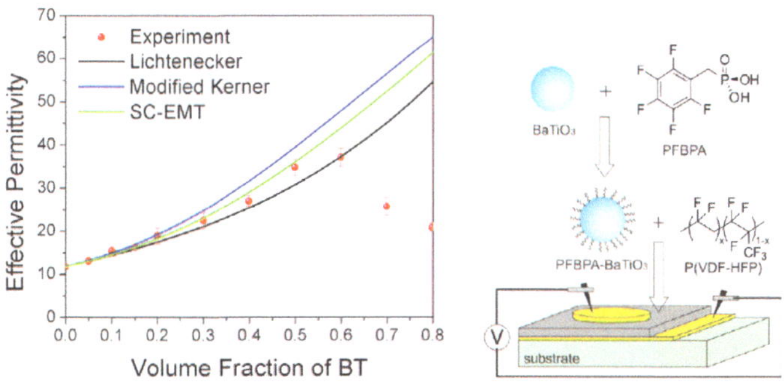

Figure 3.10 Dielectric permittivity of poly(vinylidene fluoride-co-hexafluoropropylene)/phosphonic acid modified $BaTiO_3$ nanocomposites. Reprinted with permission from Ref. [40], Copyright 2009, American Chemical Society.

Carboxylates or carboxylic acids [13, 57, 70] are commonly used surfactants, too. They can bind with $BaTiO_3$ and other oxides via chemisorption. Tetrafluorophthalic acid was used to modify the surface of $BaTiO_3$ to improve their dispersion as well as interfacial bonding with PVDF matrix [86]. The resulting PVDF/modified $BaTiO_3$ nanocomposites showed lower dielectric permittivity compared with PVDF/untreated $BaTiO_3$. However, the surface modification successfully improved the dielectric breakdown strength of the PVDF nanocomposites. The authors believed that the improved dispersion and interfacial bonding due to surface modification played a critical role in this improvement. As shown in Fig. 3.11, first of all, the uniform dispersion of nanoparticles and strong interfacial bonding as a result of surface modification could introduce more space charges in the polymer matrix; secondly, the surfactant formed a passivation layer on the particle surface, and this layer can reduce ionic or DC conduction originated from $BaTiO_3$ nanoparticles in the nanocomposites; finally, the uniform dispersion also leads to more interfacial area, which has positive influences on the interfacial charge around the surface of nanoparticles and the charge concentration associated with the diffuse double layers between nanoparticles and PVDF matrix.

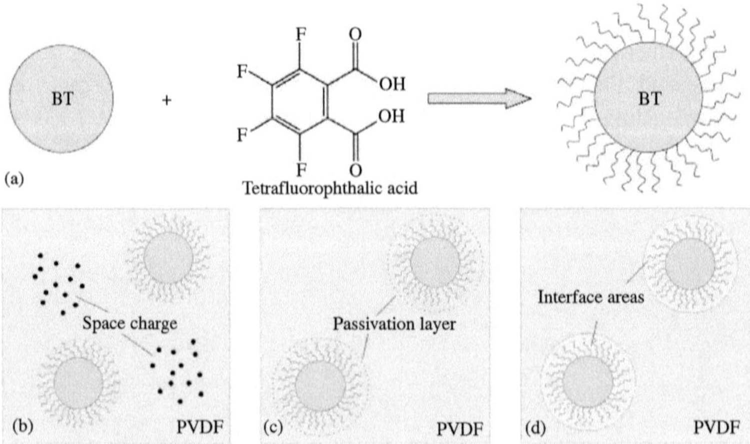

Figure 3.11 Tetrafluorophthalic acid modified $BaTiO_3$ nanoparticle (a) and its effects in PVDF nanocomposites (b–d). Reprinted with permission from Ref. [86], Copyright 2013, AIP Publishing LLC.

3.3.2 Surface Modification via Polymers

Instead of grafting small surfactant molecules onto $BaTiO_3$ surfaces, polymeric surfactants have been gaining increased attractiveness for surface modification of nanoparticles. Some of the advantages of polymers include the following: (1) The polymeric surfactant could form surface layer with different thicknesses and create a core–shell structure together with the nanoparticles. This shell layer structure can be adjusted via controlling the polymerization processes. In particular, the thickness of shell layer can be precisely controlled. (2) The long polymer chain could form more effective interaction/entanglement with polymer matrix and then lead to stronger interfacial bonding.

Figure 3.12 summarizes the strategies for creating core–shell nanostructures for polymer nanocomposites with high dielectric permittivity. The "grafting to" method directly uses polymeric surfactant to modify the particle surfaces and form a shell layer, while in the "grafting from" method, the particle surface act as initiation sites for polymerization of the monomers. With the help of atom transfer radical polymerization (ATRP) [23] and reversible addition–fragmentation chain transfer polymerization (RAFT) [95], the fabrication of core–shell structures with precisely controlled shell structure is possible now. However, the application of this "grafting from" method is limited by the fact that both polymerization techniques are only limited to synthesis of certain polymer structures. Furthermore, the core–shell structure can be created via introducing another inorganic ceramic layer onto the surface of ceramic nanoparticles. In this chapter, the all ceramic core–shell structures will be considered as a type of hybrid nanomaterials, and they will be discussed in Chapter 7, which is on hybrid nanomaterials reinforced polymer nanocomposites. It is necessary to point out that the aforementioned surface modification strategies can be used to modify other types of ceramic materials as well.

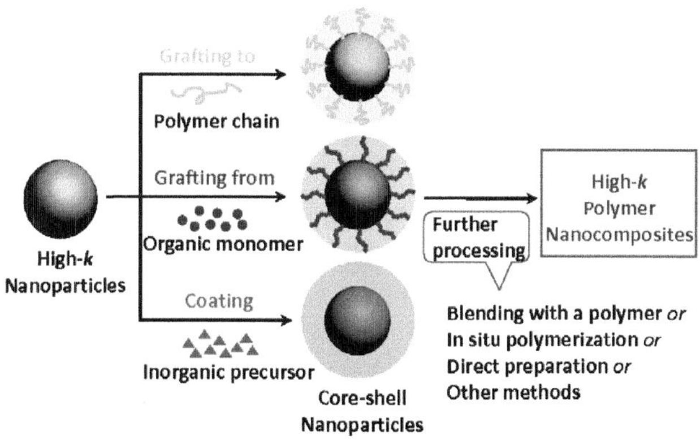

Figure 3.12 General methods associated with design and construction of core–shell nanoparticles for high-permittivity polymer nanocomposites. Reprinted from Ref. [34], with permission from John Wiley and Sons.

In the "grafting to" method, a polymer coating layer can be introduced via both (1) physical absorption as a result of electrostatic interaction and van der Waals interaction [17, 87] and (2) chemical reactions between the functional groups on particle surface and polymer coatings [76].

Polyvinylpyrrolidone (PVP) is a water-soluble polymer with high polarity. It is considered a good surfactant for PVDF/$BaTiO_3$ nanocomposites, because of its polar interactions with different groups, including the hydroxyl groups on $BaTiO_3$ particles and fluorine groups [87]. Thus, PVP can be absorbed on particle surfaces and improve the dispersion of $BaTiO_3$ nanoparticles in PVDF matrix, as well as strengthen the interfacial bonding between nanoparticles and polymer matrix. In this study (Fig. 3.13, Ref. [87]), the PVP modification successfully prevented aggregation of $BaTiO_3$ nanoparticles, and a uniform dispersion was achieved. The surface modification did not change the concentration dependence of dielectric properties. However, the PVP-modified $BaTiO_3$ nanoparticles resulted in an increase in dielectric permittivity, compared with untreated $BaTiO_3$, when the loading level is above 30 vol%, suggesting the positive contributions of PVP surface modification. In addition, the higher dielectric breakdown strength was found in the PVP-modified PVDF nanocomposites.

Polymer materials can also be coated onto the surface via chemical bonding. For instance, poly(acrylic acid) was found chemically bonded to $BaTiO_3$ surface when albumen nanocomposites were fabricated [76]. Poly(acrylic acid) has a linear polymer structure with multiple polar sites on polymer chains, and they have good compatibility with the polymer matrix (albumen) and improve the dispersion of nanoparticles and dielectric properties. Besides conventional linear polymer structures, hyperbranched polymer structure has been applied in surface modification to improve dielectric performances of polymer nanocomposites. In Fig. 3.14 [80], a hyperbranched aromatic polyamide (BT-HBP) was used to modify $BaTiO_3$ nanoparticles, and the dielectric properties of poly(vinylidene fluoride-trifluoroethylene-chlorofluoroethylene) (PVDF-TrFE-CFE) were improved. This hyperbranched surface coating resulted in both high dielectric constant and dielectric breakdown strength, which did not usually occur to conventional percolative polymer nanocomposites. It is noted that before grafting this polyamide onto particle surface, the $BaTiO_3$ nanoparticles were hydroxylated by hydrogen peroxide followed by silanization. These two steps created a large number of highly reactive amine groups for HBP modification. The resulting hyperbranched polymer coating led to remarkable improvement in dielectric constant, in particular, at high loading levels.

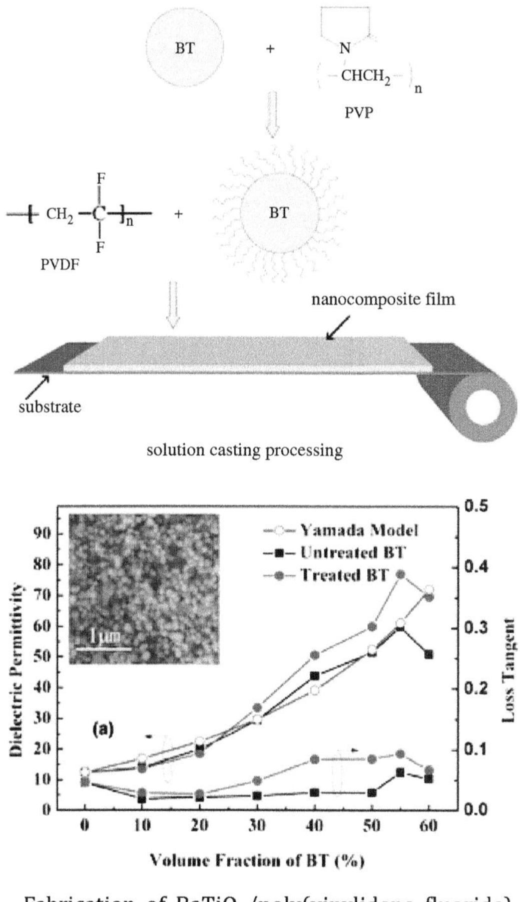

Figure 3.13 Fabrication of $BaTiO_3$/poly(vinylidene fluoride) (BT/PVDF) nanocomposite films and their dielectric properties. Reprinted from Ref. [87], with permission from John Wiley and Sons.

These studies provided solid proofs of the critical roles of polymer coatings in dielectric performances of polymer nanocomposites. As a matter of fact, design and fabrication of polymer coatings have become an effective and even powerful tool for modifying dielectric properties of polymer nanocomposites. In addition to homopolymers, copolymers have also shown to be successful in improving dielectric properties of polymer nanocomposites [77]. The same research team in Ref. [80] grafted poly(methyl methacrylate) (PMMA) on HBP-modified $BaTiO_3$ nanoparticles, creating a copolymer coating layer on the particle surface. This coating layer has a bilayer structures, due to the diblock copolymer structures. The resulting $BaTiO_3$ nanoparticles with bilayer polymer coating were incorporated into PMMA to fabricate PMMA/$BaTiO_3$ nanocomposites. As shown in Fig. 3.15, the surface modification enhanced dielectric constant of the polymer nanocomposites. Meanwhile, by comparing Figs. 3.14 and 3.15, we could find that the dielectric properties of polymer matrix had significant impacts on the dielectric properties of the polymer nanocomposites.

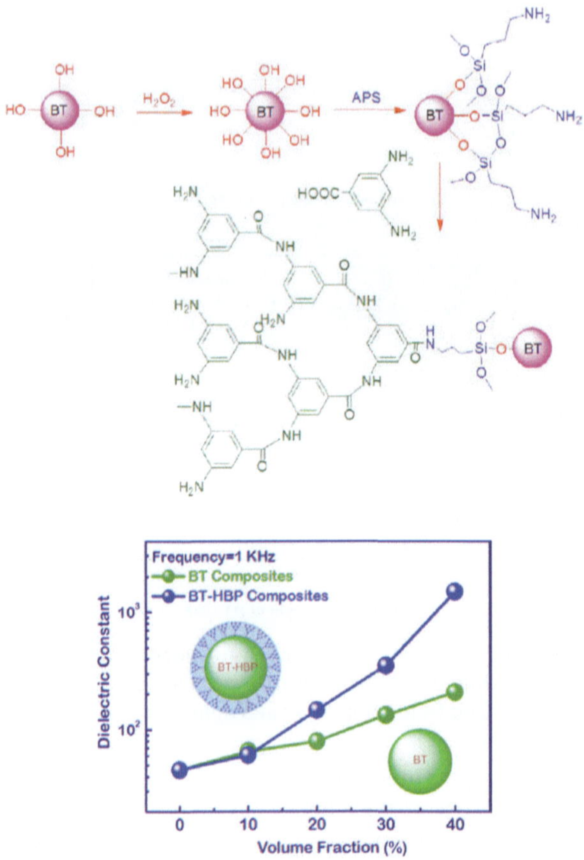

Figure 3.14 Hyperbranched aromatic polyamide grafted $BaTiO_3$ (BT-HBP) for poly(vinylidene fluoride-trifluoroethylene-chlorofluoroethylene) (PVDF-TrFE-CFE) nanocomposites and their dielectric constants. Reprinted with permission from Ref. [80], Copyright 2013, American Chemical Society.

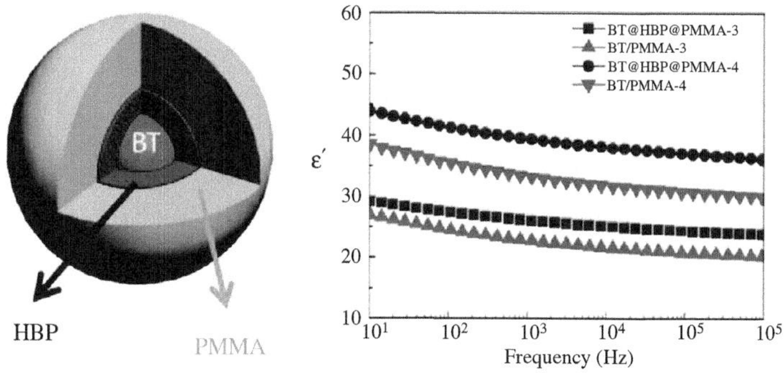

Figure 3.15 Schemes of surface modified $BaTiO_3$ nanoparticles with bilayer copolymer coating and dielectric properties of resulting polymer nanocomposites. Reprinted with permission from Ref. [77], Copyright 2013, American Chemical Society.

Another interesting example of copolymer coating is shown in Fig. 3.16. In this study, both diblock copolymer and random copolymer of polystyrene-block-poly(styrene-co-vinylbenzylchloride) (PS-b-PSVBC) reacted with NaH-modified $BaTiO_3$ nanoparticles, and formed a coating layer chemically bonded to $BaTiO_3$ surface. This surface coating was further modified by trimethylamine (Et_3N) to bury the charged layers within the coating layer and minimize leakage current. The reaction mechanisms between copolymers and NaH-modified $BaTiO_3$ nanoparticles are illustrated in Fig. 3.16. The random copolymer (r-PSVBC) was also used to modify $BaTiO_3$ nanoparticles via the same procedures as PS-b-PSVBC for comparative studies. The modified nanoparticles were incorporated into polystyrene to make polymer nanocomposites. Besides the remarkable concentration dependence, the copolymer structures also had great impacts on the dielectric permittivity. According to Fig. 3.17, the diblock copolymer (PS-b-PSVBC) favors higher dielectric permittivity and breakdown strength, compared with random copolymer (r-PSVBC).

The polymer coating can be introduced by "grafting from" method as well. In this method, synthesis of the polymer coating is initiated on the surface of nanoparticles. In particular, with rapid development of some new polymer synthesis techniques, such as atom transfer radical polymerization (ATRP) [62, 67, 91] and reversible addition–fragmentation chain transfer polymerization (RAFT) [44, 55, 59, 83,

91], this "grafting from" approach has been popularized in making high–dielectric performance polymer nanocomposites. Both ATRP and RAFT are controlled free radical polymerization. Compared with conventional free radical polymerization, they could produce polymer structures with precisely controlled molecular weight, and narrow molecular weight distribution. In the meantime, they are also very powerful in synthesis of complex polymer structures, including controlled block copolymers and hyperbranched polymer structures. These new polymer synthesis techniques with precise control over polymer structures create great opportunities for the design and customization of the composition and structures of the surface coatings, which in turn will be able to tune interfacial polarization in polymer nanocomposites [59]. On the other hand, due to their precise control over polymer structures, they also allow people to achieve an in-depth understanding of the roles of the interface between polymer matrix and nanoparticles in dielectric polarization and related properties.

For example, the study shown in Fig. 3.18 used surface initiated RAFT to synthesize different polymer coating layers on the surface of $BaTiO_3$ nanoparticles. First of all, the $BaTiO_3$ nanoparticles were hydroxylated in hydrogen peroxide followed by silanization. The silanization introduced reactive amino groups ($-NH_2$) onto the particles surface, which could react with RAFT agent S-1-dodecyl-S′-(α, α'-dimethyl-α''-acetic acid) trithiocarbonate (DDMAT). Three different types of polymer coating layers were synthesized on the DDMAT-modified $BaTiO_3$. The same thickness of different polymer coating layers were conveniently realized during RAFT polymerization by controlling the molecular weight. This helps understand the roles of different polymer coating structures in the interfacial polarization in polymer nanocomposites. It was found that the electrical properties of the polymer coating layer was critical to dielectric breakdown strength. To achieve higher breakdown strength, the polymer coating with lower electrical conductivity should be applied to the nanoparticles. Certainly, the thickness of the polymer coating layer is important to the dielectric properties too. In a study on the thiophene polymer-modified $BaTiO_3$ /polymer nanocomposites, the layer thickness was again adjusted by controlling its molecular weight. It turned out the thickness could lead to remarkable difference in energy density of the polymer nanocomposites. When the thickness increased from approximately 8 nm to 15 nm, the increase in stored energy density was observed in the nanocomposites [59]. The effects of thickness of coating layers on dielectric properties were also reported in fluoropolymer-modified $BaTiO_3$/polymer nanocomposites [83]. The surface initiated RAFT was also used to synthesize polystyrene/$BaTiO_3$ nanocomposites. The PS did not only act as the coating layer, but also act as a matrix. According to the authors, both high dielectric constant (~25) and low dielectric loss (<0.03) were achieved in the nanocomposites [82].

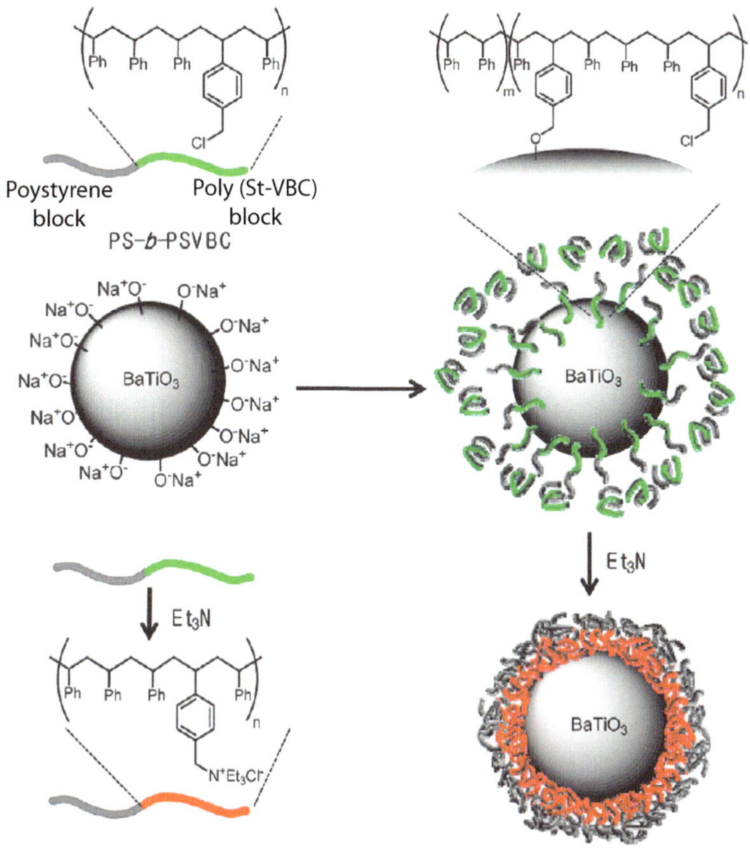

Figure 3.16 Scheme of $BaTiO_3$ nanoparticles with polystyrene-block-poly(styrene-co-vinylbenzylchloride) (PS-b-PSVBC) diblock copolymer coating. Reprinted with permission from Ref. [37], Copyright 2010, American Chemical Society.

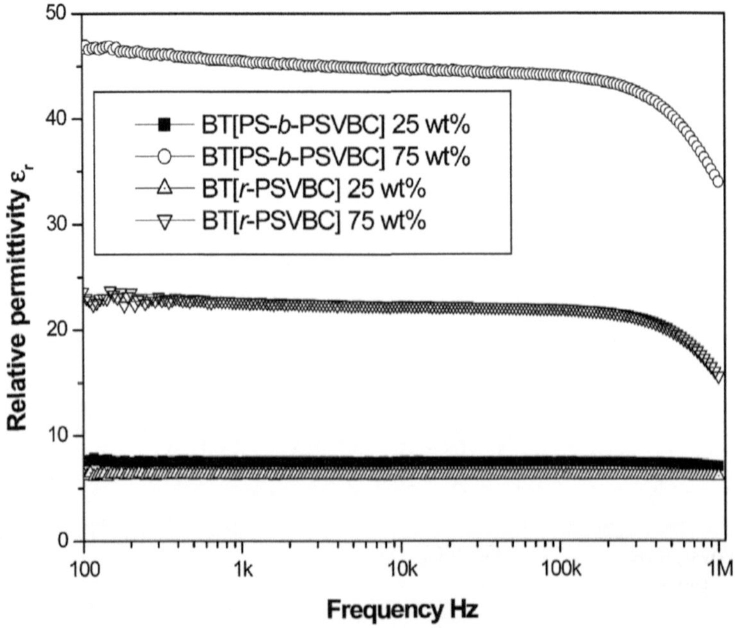

Figure 3.17 Dielectric permittivity of BT[PS-b-PSVBC]/PS and BT[r-PSVBC]/PS nanocomposite thin films. Reprinted with permission from Ref. [37], Copyright 2010, American Chemical Society.

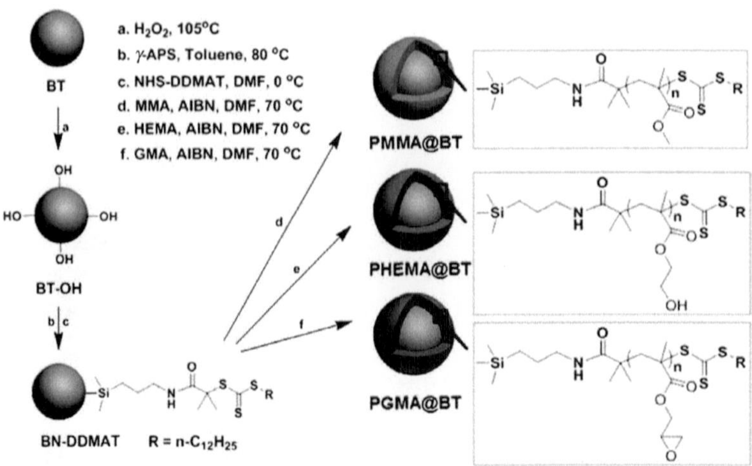

Figure 3.18 Scheme illustrating the preparation process of the polymer@ BT nanoparticles by surface-initiated RAFT polymerization. Reprinted with permission from Ref. [95], Copyright 2014, American Chemical Society.

Another popular controlled free-radical polymerization is atom transfer radical polymerization (ATRP) [75, 78, 79, 81, 88]. Similar to RAFT, the ATRP agent (initiator) should be grafted onto the particles surface first, in order to initiate ATRP, such as the organic halides (Fig. 3.19). Meanwhile, copper (CuCl and $CuCl_2$) as catalyst is also critical to ATRP. In the study shown in Fig. 3.19. The poly(glycidyl methacrylate) was grafted onto the surface via surface initiated ATRP. The resulting coating layer did not only improve dispersion, but also provide sufficient mechanical stability. Meanwhile, enhancement of dielectric permittivity was observed, due to improved interfacial polarization. In poly(1H,1H,2H,2H-perfluorooctyl methacrylate) (PPFOMA)@$BaTiO_3$ nanocomposites, the PPFOMA was coated on the nanoparticle surface via surface initiated ATRP. The existence of this polymer coating layer resulted in both improved dielectric constant and reduced dielectric loss. In particular, the dielectric loss of the nanocomposites was much lower than that of pure polymers, which usually did not occur in dielectric polymer nanocomposites. Meanwhile, this work also found the increase of PPFOMA/$BaTiO_3$ (w/w) ratio could lead to the decrease in dielectric constant. This finding further proved the significant, but complex effects of surface modification of nanoparticles on interfacial polarization in the nanocomposites [88].

The chemical structures of ATRP initiator is another important structural factor in interfacial polarization [75]. In a comparative study between a highly polarizable initiator (tetrameric metallophthalocyanine (TMPc) type initiator) (Fig. 3.20) and a less polarizable initiator (phosphonate type initiator (R2)) (Fig. 3.21), both initiators were used to synthesize poly(methyl methacrylate) (PMMA) on the surface of $BaTiO_3$ nanoparticles. The interfacial layer formed by this highly polarizable TMPc-Br initiator was believed to contribute to the higher dielectric energy storage capability in the resulting poly(vinylidene fluoride-co-hexafluoropropylene) (P(VDF-HFP)) nanocomposites.

These studies suggested that polymer coating on nanoparticles is effective to modify the dielectric polarization in polymer nanocomposites. In particular, with the assistance of controlled polymerization technique, the precisely designed layer structures with desired properties can be conveniently created. However, there are some issues we have to face. For example, the controlled polymerization technique could only be applied for synthesis of limited types of polymers, typically, PMMA and PS. This is why these two polymers have been frequently used to modify nanoparticles using surface initiated ATRP and RAFT techniques [58, 78, 79, 81]. Meanwhile, in RAFT polymerization, the RAFT agents are not very durable. Over time, it will gradually decompose. In order to explore their full potential, these issues need to be resolved.

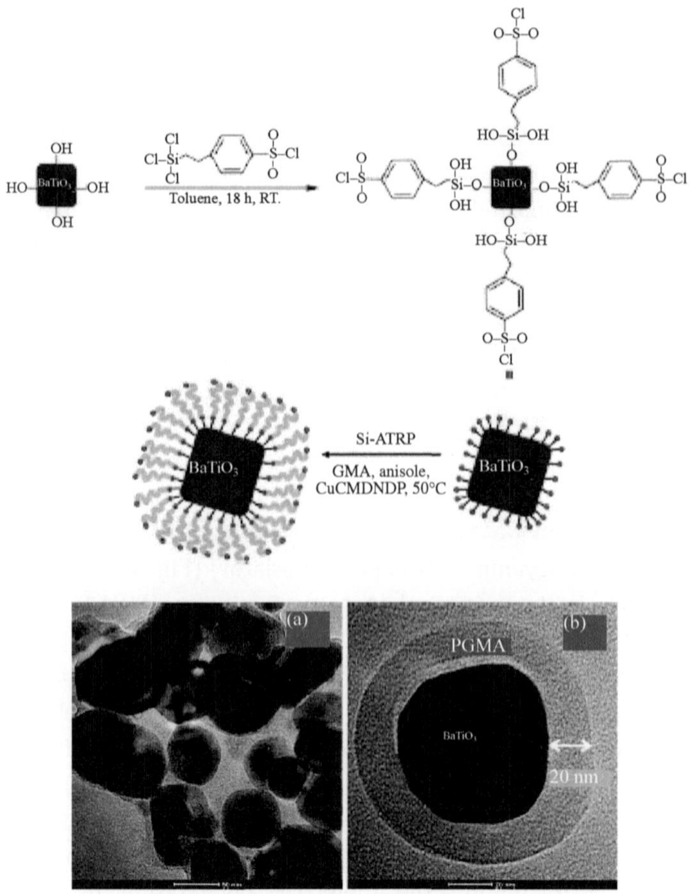

Figure 3.19 Synthesis of $BaTiO_3$-PGMA core–shell nanocomposites by Si-ATRP of GMA (glycidyl methacrylate) from $BaTiO_3$ nanoparticles. Reprinted from Ref. [23], with permission from John Wiley and Sons.

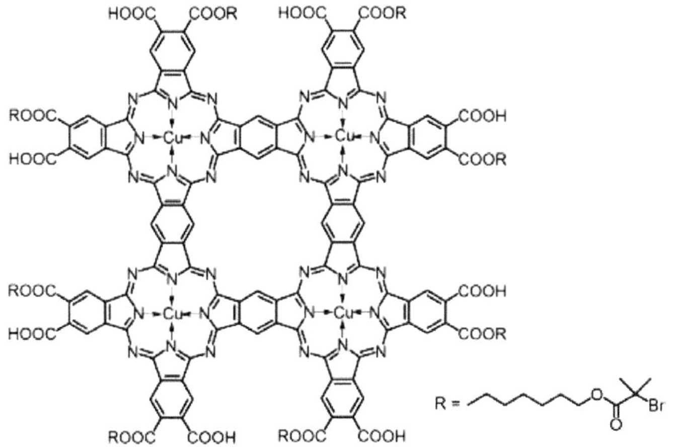

Figure 3.20 Chemical structure of the TMPc-Br ATRP initiator. Reprinted from Ref. [75], with permission from John Wiley and Sons.

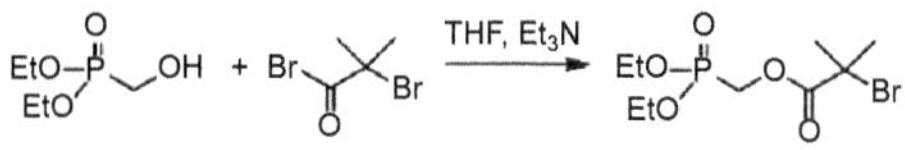

Figure 3.21 Synthesis of the R2-Br ATRP initiator. Reprinted from Ref. [75], with permission from John Wiley and Sons.

References

1. Abd-El Messieh, S. L. and Rozik, N. N. (2011). Dielectric and morphological studies on polyester/nanosilica fume composites. *J Appl Polym Sci*, **122**:714–721.

2. Abrar, A., Zhang, D., Su, B., Button, T. W., Kirk, K. J., and Cochran, S. (2004). 1-3 connectivity piezoelectric ceramic-polymer composite transducers made with viscous polymer processing for high frequency ultrasound. *Ultrasonics*, **42**:479–484.

3. Adam, J., Lehnert, T., Klein, G., and McMeeking, R. M. (2014). Ferroelectric properties of composites containing $BaTiO_3$ nanoparticles of various sizes. *Nanotechnology*, **25**.

4. Adireddy, S., Puli, V. S., Lou, T. J., Elupula, R., Sklare, S. C., Riggs, B. C., and Chrisey, D. B. (2015). Polymer-ceramic nanocomposites for high energy density applications. *J Sol-Gel Sci Technol*, **73**:641–646.

5. Aftab, S., Hall, D. A., Aleem, M. A., and Siddiq, M. (2013). Low field ac study of PZT/PVDF nano composites. *J Mater Sci-Mater El*, **24**:979–986.

6. Ahmad, G., Dickerson, M. B., Cai, Y., Jones, S. E., Ernst, E. M., Vernon, J. P., Haluska, M. S., Fang, Y., Wang, J., Subrarnanyarn, G., Naik, R. R., and Sandhage, K. H. (2008). Rapid bioenabled formation of ferroelectric $BaTiO_3$ at room temperature from an aqueous salt solution at near neutral pH. *J Am Chem Soc*, **130**:4–5.

7. Al Jaafari, A. A. and Ayesh, A. S. (2011). Effect of ZnO nano-particles on the dielectric relaxation behavior and thermal stability of polycarbonate host. *J Thermoplast Compos Mater*, **24**:837–852.

8. Almadhoun, M. N., Bhansali, U. S., and Alshareef, H. N. (2012). Nanocomposites of ferroelectric polymers with surface-hydroxylated $BaTiO_3$ nanoparticles for energy storage applications. *J Mater Chem*, **22**:11196–11200.

9. Arjmand, M., Mahmoodi, M., Park, S., and Sundararaj, U. (2013). An innovative method to reduce the energy loss of conductive filler/polymer composites for charge storage applications. *Compos Sci Technol*, **78**:24–29.

10. Arlt, K. and Wegener, M. (2010). Piezoelectric PZT/PVDF-copolymer 0-3 composites: aspects on film preparation and electrical poling. *IEEE Trans Dielectr Electr Insul*, **17**:1178–1184.

11. Assink, R. A. and Schwartz, R. W. (1993). H-1 and C-13 NMR investigations of $Pb(Zr,Ti)O_3$ thin-film precursor solutions. *Chem Mater*, **5**:511–517.

12. Begg, B. D., Vance, E. R., and Nowotny, J. (1994). Effect of particle-size on the room-temperature crystal-structure of barium-titanate. *J Am Ceram Soc*, **77**:3186–3192.

13. Beier, C. W., Cuevas, M. A., and Brutchey, R. L. (2010). Effect of surface modification on the dielectric properties of $BaTiO_3$ nanocrystals. *Langmuir*, **26**:5067–5071.

14. Choudhury, A. (2012). Preparation, characterization and dielectric properties of polyetherimide nanocomposites containing surface-functionalized $BaTiO_3$ nanoparticles. *Polym Int*, **61**:696–702.

15. Cong, R., Hu, G. J., Yu, G. L., Chu, J. H., and Dai, N. (2015). Effect of polymer content on the optical properties of PbZr0.4 Ti-0.6 O-3 multilayers. *J Infrared Millimeter Waves*, **34**:297–300.

16. Culver, S. P., Beier, C. W., Rafson, J. P., and Brutchey, R. L. (2014). Surface modification of $BaTiO_3$ inclusions in polydicyclopentadiene nanocomposites for energy storage. *J Appl Polym Sci*, **131**.

17. Dang, Z. M., Lin, Y. Q., Xu, H. P., Shi, C. Y., Li, S. T., and Bai, J. B. (2008). Fabrication and dielectric characterization of advanced $BaTiO_3$/

polyimide nanocomposite films with high thermal stability. *Adv Funct Mater*, **18**:1509–1517.

18. Dang, Z. M., Wang, H. Y., Peng, B., and Nan, C. W. (2008). Effect of $BaTiO_3$ size on dielectric property of $BaTiO_3$/PVDF composites. *J Electroceram*, **21**:381–384.

19. Dang, Z. M., Yuan, J. K., Yao, S. H., and Liao, R. J. (2013). Flexible nanodielectric materials with high permittivity for power energy storage. *Adv Mater*, **25**:6334–6365.

20. Ding, Y. C., Wu, Q., Zhao, D., Ye, W., Hanif, M., and Hou, H. Q. (2013). Flexible PI/$BaTiO_3$ dielectric nanocomposite fabricated by combining electrospinning and electrospraying. *Eur Polym J*, **49**:2567–2571.

21. Dou, X., Liu, X., Zhang, Y., Feng, H., Chen, J.-F., and Du, S. (2009). Improved dielectric strength of barium titanate-poly(vinylidene fluoride) nanocomposite. *Appl Phys Lett*, **95**:132904.

22. Ehrhardt, C., Fettkenhauer, C., Glenneberg, J., Munchgesang, W., Pientschke, C., Grossmann, T., Zenkner, M., Wagner, G., Leipner, H. S., Buchsteiner, A., Diestelhorst, M., Lemm, S., Beige, H., and Ebbinghaus, S. G. (2013). $BaTiO_3$-P(VDF-HFP) nanocomposite dielectrics—Influence of surface modification and dispersion additives. *Mater Sci Eng B-Adv*, **178**:881–888.

23. Ejaz, M., Puli, V. S., Elupula, R., Adireddy, S., Riggs, B. C., Chrisey, D. B., and Grayson, S. M. (2015). Core-shell structured poly(glycidyl methacrylate)/$BaTiO_3$ nanocomposites prepared by surface-initiated atom transfer radical polymerization: a novel material for high energy density dielectric storage. *J Polym Sci Pol Chem*, **53**:719–728.

24. Fan, B. H., Zha, J. W., Wang, D. R., Zhao, J., and Dang, Z. M. (2012). Experimental study and theoretical prediction of dielectric permittivity in $BaTiO_3$/polyimide nanocomposite films. *Appl Phys Lett*, **100**.

25. Fan, B. H., Zha, J. W., Wang, D. R., Zhao, J., and Dang, Z. M. (2012). Size-dependent low-frequency dielectric properties in the $BaTiO_3$/poly(vinylidene fluoride) nanocomposite films. *Appl Phys Lett*, **100**.

26. Fan, B. H., Zha, J. W., Wang, D. R., Zhao, J., Zhang, Z. F., and Dang, Z. M. (2013). Preparation and dielectric behaviors of thermoplastic and thermosetting polymer nanocomposite films containing $BaTiO_3$ nanoparticles with different diameters. *Compos Sci Technol*, **80**:66–72.

27. Frey, M. H. and Payne, D. A. (1996). Grain-size effect on structure and phase transformations for barium titanate. *Phys Rev B*, **54**:3158–3168.

28. Gao, L., He, J. L., Hu, J., and Li, Y. (2014). Large enhancement in polarization response and energy storage properties of poly(vinylidene

fluoride) by improving the interface effect in nanocomposites. *J Phys Chem C*, **118**:831–838.

29. Gevorkyan, A., Shter, G. E., Shmueli, Y., Buk, A., Meir, R., and Grader, G. S. (2014). Branching effect and morphology control in electrospun PbZr0.52Ti0.48O3 nanofibers. *J Mater Res*, **29**:1721–1729.

30. Hamciuc, E., Hamciuc, C., Bacosca, I., Cristea, M., and Okrasa, L. (2011). Thermal and electrical properties of nitrile-containing polyimide/$BaTiO_3$ composite films. *Polym Compos*, **32**:846–855.

31. Hsiang, H. I., Lin, K. Y., Yen, F. S., and Hwang, C. Y. (2001). Effects of particle size of $BaTiO_3$ powder on the dielectric properties of $BaTiO_3$/poly(vinylidene fluoride) composites. *J Mater Sci*, **36**:3809–3815.

32. Huang, X., Xie, L., Yang, K., Wu, C., Jiang, P., Li, S., Wu, S., Tatsumi, K., and Tanaka, T. (2014). Role of interface in highly filled epoxy/$BaTiO_3$ nanocomposites. Part II: Effect of nanoparticle surface chemistry on processing, thermal expansion, energy storage and breakdown strength of the nanocomposites. *IEEE Trans Dielectr Electr Insul*, **21**:480–487.

33. Huang, X. Y., Xie, L. Y., Yang, K., Wu, C., Jiang, P. K., Li, S. T., Wu, S., Tatsumi, K., and Tanaka, T. (2014). Role of interface in highly filled epoxy/$BaTiO_3$ nanocomposites. Part I: Correlation between nanoparticle surface chemistry and nanocomposite dielectric property. *IEEE Trans Dielectr Electr Insul*, **21**:467–479.

34. Huang, X. Y., and Jiang, P. K. (2015). Core-shell structured high-k polymer nanocomposites for energy storage and dielectric applications. *Adv Mater*, **27**:546–554.

35. Jain, A., Prashanth, K. J., Sharma, A. K., Jain, A., and Rashmi, P. N. (2015). Dielectric and piezoelectric properties of PVDF/PZT composites: a review. *Polym Eng Sci*, **55**:1589–1616.

36. Joshi, U. A., Yoon, S. H., Baik, S. G., and Lee, J. S. (2006). Surfactant-free hydrothermal synthesis of highly tetragonal barium titanate nanowires: a structural investigation. *J Phys Chem B*, **110**:12249–12256.

37. Jung, H. M., Kang, J.-H., Yang, S. Y., Won, J. C., and Kim, Y. S. (2010). Barium titanate nanoparticles with diblock copolymer shielding layers for high-energy density nanocomposites. *Chem Mater*, **22**:450–456.

38. Kamezawa, N., Nagao, D., Ishii, H., and Konno, M. (2015). Transparent, highly dielectric poly(vinylidene fluoride) nanocomposite film homogeneously incorporating $BaTiO_3$ nanoparticles with fluoroalkylsilane surface modifier. *Eur Polym J*, **66**:528–532.

39. Kim, P., Zhang, X. H., Domercq, B., Jones, S. C., Hotchkiss, P. J., Marder, S. R., Kippelen, B., and Perry, J. W. (2008). Solution-processible high-permittivity nanocomposite gate insulators for organic field-effect transistors. *Appl Phys Lett*, **93**:013302.

40. Kim, P., Doss, N. M., Tillotson, J. P., Hotchkiss, P. J., Pan, M. J., Marder, S. R., Li, J. Y., Calame, J. P., and Perry, J. W. (2009). High energy density nanocomposites based on surface-modified $BaTiO_3$ and a ferroelectric polymer. *ACS Nano*, **3**:2581–2592.

41. Kim, Y. S., Smith, O. L., Kathaperumal, M., Johnstone, L. R., Pan, M. J., and Perry, J. W. (2014). Enhancement of breakdown strength and energy density in $BaTiO_3$/ferroelectric polymer nanocomposites via processing-induced matrix crystallinity and uniformity. *RSC Adv*, **4**:19668–19673.

42. Klicker, K. A., Biggers, J. V., and Schulze, W. A. (1980). Effect of polymer matrix on hydrostatic response of Pzt-polymer transducers. *Am Ceram Soc Bull*, **59**:381–381.

43. Kobayashi, Y., Nishikata, A., Tanase, T., and Konno, M. (2004). Size effect on crystal structures of barium titanate nanoparticles prepared by a sol-gel method. *J Sol-Gel Sci Technol*, **29**:49–55.

44. Kwak, Y., Nicolay, R., and Matyjaszewski, K. (2008). Concurrent ATRP/RAFT of styrene and methyl methacrylate with dithioesters catalyzed by copper(I) complexes. *Macromolecules*, **41**:6602–6604.

45. Lai, M. B., Yu, S. H., Sun, R., Zeng, X. L., Luo, S. B., and Wong, C. P. (2013). Effects and mechanism of graft modification on the dielectric performance of polymer-matrix composites. *Compos Sci Technol*, **89**:127–133.

46. Li, Q., Han, K., Gadinski, M. R., Zhang, G. Z., and Wang, Q. (2014). High energy and power density capacitors from solution-processed ternary ferroelectric polymer nanocomposites. *Adv Mater*, **26**:6244–6249.

47. Liu, S. H., Xue, S. X., Zhang, W. Q., and Zhai, J. W. (2014). Enhanced dielectric and energy storage density induced by surface-modified $BaTiO_3$ nanofibers in poly(vinylidene fluoride) nanocomposites. *Ceram Int*, **40**:15633–15640.

48. Luo, H., Zhang, D., Jiang, C., Yuan, X., Chen, C., and Zhou, K. C. (2015). Improved dielectric properties and energy storage density of poly(vinylidene fluoride-co-hexafluoropropylene) nanocomposite with hydantoin epoxy resin coated $BaTiO_3$. *ACS Appl Mater Interfaces*, **7**:8061–8069.

49. Mao, Y. P., Mao, S. Y., Ye, Z. G., Xie, Z. X., and Zheng, L. S. (2010). Size-dependences of the dielectric and ferroelectric properties of $BaTiO_3$/poly(vinylidene fluoride) nanocomposites. *J Appl Phys*, **108**.

50. Mendes, S. F., Costa, C. M., Caparros, C., Sencadas, V., and Lanceros-Mendez, S. (2012). Effect of filler size and concentration on the structure and properties of poly(vinylidene fluoride)/$BaTiO_3$ nanocomposites. *J Mater Sci*, **47**:1378–1388.

51. Mirza, M. S., Yasin, T., Ikram, M., Khan, M. N., and Shuaib, M. (2014). Fabrication of 1–3 connectivity (Ba0.95Pb0.05)(Ti0.99Co0.01)O-3/monothane-A70 composites by die-pressing method and their electrical characterizations. *Ceram Int*, **40**:11477–11484.

52. Modeshia, D. R. and Walton, R. I. (2010). Solvothermal synthesis of perovskites and pyrochlores: crystallisation of functional oxides under mild conditions. *Chem Soc Rev*, **39**:4303–4325.

53. Mofokeng, T. G., Luyt, A. S., Pavlovic, V. P., Pavlovic, V. B., Dudic, D., Vlahovic, B., and Djokovic, V. (2014). Ferroelectric nanocomposites of poly(vinylidene fluoride)/polymethyl methacrylate blend and $BaTiO_3$ particles: fabrication of beta-crystal polymorph rich matrix through mechanical activation of the filler. *J Appl Phys*, **115**.

54. Nagao, D., Kinoshita, T., Watanabe, A., and Konno, M. (2011). Fabrication of highly refractive, transparent $BaTiO_3$/poly(methyl methacrylate) composite films with high permittivities. *Polym Int*, **60**:1180–1184.

55. Nicolay, R., Kwak, Y., and Matyjaszewski, K. (2008). Dibromotrithiocarbonate iniferter for concurrent ATRP and RAFT polymerization. Effect of monomer, catalyst, and chain transfer agent structure on the polymerization mechanism. *Macromolecules*, **41**:4585–4596.

56. Niu, Y. J., Yu, K., Bai, Y. Y., and Wang, H. (2015). Enhanced dielectric performance of $BaTiO_3$/PVDF composites prepared by modified process for energy storage applications. *IEEE Trans Ultrason Ferroelectr Freq Control*, **62**:108–115.

57. O'Brien, S., Brus, L., and Murray, C. B. (2001). Synthesis of monodisperse nanoparticles of barium titanate: toward a generalized strategy of oxide nanoparticle synthesis. *J Am Chem Soc*, **123**:12085–12086.

58. Paniagua, S. A., Kim, Y., Henry, K., Kumar, R., Perry, J. W., and Marder, S. R. (2014). Surface-initiated polymerization from barium titanate nanoparticles for hybrid dielectric capacitors. *ACS Appl Mater Interfaces*, **6**:3477–3482.

59. Qiao, Y. L., Islam, M. S., Wang, L., Yan, Y., Zhang, J. Y., Benicewicz, B. C., Ploehn, H. J., and Tang, C. B. (2014). Thiophene polymer-grafted barium titanate nanoparticles toward nanodielectric composites. *Chem Mater*, **26**:5319–5326.

60. Rabuffetti, F. A. and Brutchey, R. L. (2013). Local structure of $Ba_{1-x}Sr_xTiO_3$ and $BaTi_{1-y}Zr_yO_3$ nanocrystals probed by X-ray absorption and X-ray total scattering. *ACS Nano*, **7**:11435–11444.

61. Rabuffetti, F. A. and Brutchey, R. L. (2014). Complex perovskite oxide nanocrystals: low-temperature synthesis and crystal structure. *J Chem Soc Dalton T*, **43**:14499–14513.

62. Ran, J., Wu, L., Zhang, Z. H., and Xu, T. W. (2014). Atom transfer radical polymerization (ATRP): a versatile and forceful tool for functional membranes. *Prog Polym Sci*, **39**:124–144.

63. Shrout, T. R., Schulze, W. A., and Biggers, J. V. (1979). Simplified fabrication of PZT-polymer composites. *Mater Res Bull*, **14**:1553–1559.

64. Shrout, T. R., Bowen, L. J., and Schulze, W. A. (1980). Extruded PZT-polymer composites for electro-mechanical transducer applications. *Mater Res Bull*, **15**:1371–1379.

65. Shrout, T. R., Rittenmyer, K., Schulze, W. A., and Biggers, J. V. (1980). Simplified fabrication of PZT-polymer composites. *Am Ceram Soc Bull*, **59**:381–381.

66. Siddabattuni, S., Schuman, T. P., and Dogan, F. (2013). Dielectric properties of polymer-particle nanocomposites influenced by electronic nature of filler surfaces. *ACS Appl Mater Interfaces*, **5**:1917–1927.

67. Siegwart, D. J., Oh, J. K., and Matyjaszewski, K. (2012). ATRP in the design of functional materials for biomedical applications. *Prog Polym Sci*, **37**:18–37.

68. Smith, M. B., Page, K., Siegrist, T., Redmond, P. L., Walter, E. C., Seshadri, R., Brus, L. E., and Steigerwald, M. L. (2008). Crystal structure and the paraelectric-to-ferroelectric phase transition of nanoscale $BaTiO_3$. *J Am Chem Soc*, **130**:6955–6963.

69. Tsurumi, T., Hoshina, T., Takeda, H., Mizuno, Y., and Chazono, H. (2009). Size effect of barium titanate and computer-aided design of multilayered ceramic capacitors. *IEEE Trans Ultrason Ferroelectr Freq Control*, **56**:1513–1522.

70. Urban, J. J., Yun, W. S., Gu, Q., and Park, H. (2002). Synthesis of single-crystalline perovskite nanorods composed of barium titanate and strontium titanate. *J Am Chem Soc*, **124**:1186–1187.

71. Wada, S., Suzuki, T., and Noma, T. (1996). Role of lattice defects in the size effect of barium titanate fine particles: a new model. *J Ceram Soc Jpn*, **104**:383–392.

72. Wada, S., Hoshina, T., Yasuno, H., Nam, S. M., Kakemoto, H., Tsurumi, T., and Yashima, M. (2005). Size dependence of dielectric properties for nm-sized barium titanate crystallites and its origin. *J Korean Phys Soc*, **46**:303–307.

73. Wada, S., Hoshina, T., Yasuno, H., Ohishi, M., Kakemoto, H., Tsurumi, T., and Yashima, M. (2006). Size effect of dielectric properties for barium titanate particles and its model. *Electroceram Jpn VIII*, **301**:27–30.

74. Wada, S., Hoshina, T., Takizawa, K., Ohishi, M., Yasuno, H., Kakemoto, H., Tsurumi, T., Moriyoshi, C., and Kuroiwa, Y. (2007). Origin of ultrahigh dielectric constants for barium titanate nanoparticles. *J Korean Phys Soc*, **51**:878–881.

75. Wang, J., Guan, F. X., Cui, L., Pan, J. L., Wang, Q., and Zhu, L. (2014). Achieving high electric energy storage in a polymer nanocomposite at low filling ratios using a highly polarizable phthalocyanine interphase. *J Polym Sci Pol Phys*, **52**:1669–1680.

76. Wu, X. Z., Chen, Z., and Cui, Z. (2013). Investigation of solution processable albumen-$BaTiO_3$ nanocomposite and its application in high-k films. *Compos Sci Technol*, **81**:48–53.

77. Xie, L., Huang, X., Huang, Y., Yang, K., and Jiang, P. (2013). Core@double-shell structured $BaTiO_3$–polymer nanocomposites with high dielectric constant and low dielectric loss for energy storage application. *J Phys Chem C*, **117**:22525–22537.

78. Xie, L. Y., Huang, X. Y., and Jiang, P. K. (2011). Low loss and high dielectric constant poly(methyl methacrylate)/$BaTiO_3$ nanocomposites prepared by in situ atom transfer radical polymerization. *Proceedings of 2011 International Conference on Electrical Insulating Materials (Iseim)*, pp. 92–94.

79. Xie, L. Y., Huang, X. Y., Wu, C., and Jiang, P. K. (2011). Core-shell structured poly(methyl methacrylate)/$BaTiO_3$ nanocomposites prepared by in situ atom transfer radical polymerization: a route to high dielectric constant materials with the inherent low loss of the base polymer. *J Mater Chem*, **21**:5897–5906.

80. Xie, L. Y., Huang, X. Y., Huang, Y. H., Yang, K., and Jiang, P. K. (2013). Core-shell structured hyperbranched aromatic polyamide/$BaTiO_3$ hybrid filler for poly(vinylidene fluoride-trifluoroethylene-chlorofluoroethylene) nanocomposites with the dielectric constant

comparable to that of percolative composites. *ACS Appl Mater Interfaces*, **5**:1747–1756.

81. Xie, L. Y., Huang, X. Y., Yang, K., Li, S. T., and Jiang, P. K. (2014). "Grafting to" route to PVDF-HFP-GMA/$BaTiO_3$ nanocomposites with high dielectric constant and high thermal conductivity for energy storage and thermal management applications. *J Mater Chem A*, **2**:5244–5251.

82. Yang, K., Huang, X. Y., Xie, L. Y., Wu, C., Jiang, P. K., and Tanaka, T. (2012). Core-shell structured polystyrene/$BaTiO_3$ hybrid nanodielectrics prepared by in situ RAFT polymerization: a route to high dielectric constant and low loss materials with weak frequency dependence. *Macromol Rapid Comm*, **33**:1921–1926.

83. Yang, K., Huang, X. Y., Huang, Y. H., Xie, L. Y., and Jiang, P. K. (2013). Fluoro-polymer@$BaTiO_3$ hybrid nanoparticles prepared via RAFT polymerization: toward ferroelectric polymer nanocomposites with high dielectric constant and low dielectric loss for energy storage application. *Chem Mater*, **25**:2327–2338.

84. Ye, H. J., Shao, W. Z., and Zhen, L. (2013). Tetradecylphosphonic acid modified $BaTiO_3$ nanoparticles and its nanocomposite. *Colloid Surface A*, **427**:19–25.

85. Yoon, D. H. and Lee, B. I. (2002). BaTiO(3) properties and powder characteristics for ceramic capacitors. *J Ceram Process Res*, **3**:41–47.

86. Yu, K., Niu, Y. J., Xiang, F., Zhou, Y. C., Bai, Y. Y., and Wang, H. (2013). Enhanced electric breakdown strength and high energy density of barium titanate filled polymer nanocomposites. *J Appl Phys*, **114**.

87. Yu, K., Niu, Y. J., Zhou, Y. C., Bai, Y. Y., and Wang, H. (2013). Nanocomposites of surface-modified $BaTiO_3$ nanoparticles filled ferroelectric polymer with enhanced energy density. *J Am Ceram Soc*, **96**:2519–2524.

88. Zhang, X. H., Chen, H. C., Ma, Y. H., Zhao, C. W., and Yang, W. T. (2013). Preparation and dielectric properties of core-shell structural composites of poly(1H,1H,2H,2H-perfluorooctyl methacrylate)@ $BaTiO_3$ nanoparticles. *Appl Surf Sci*, **277**:121–127.

89. Zhang, Y. Y., Li, L. P., Wang, B., Zhang, J. P., and Wang, E. P. (2014). Influence of coupling agent on microcosmic morphology and dielectric properties of energy-storage nanocomposite. *J Mater Sci-Mater El*, **25**:805–810.

90. Zhang, Y. Y., Zhang, J. P., Gao, J. X., Li, H., Wang, E. P., and Hu, X. (2015). Effect of carbon nanotubes addition on properties of 0-3 pyroelectric ceramic/polymer composites. *J Electroceram*, **34**:216–220.

91. Zhao, Y. L., Wang, L., Xiao, A. G., and Yu, H. J. (2010). The synthesis of modified polyethylene via coordination polymerization followed by ATRP, RAFT, NMRP or ROP. *Prog Polym Sci*, **35**:1195–1216.

92. Zhao, Z., Buscaglia, V., Viviani, M., Buscaglia, M. T., Mitoseriu, L., Testino, A., Nygren, M., Johnsson, M., and Nanni, P. (2004). Grain-size effects on the ferroelectric behavior of dense nanocrystalline $BaTiO_3$ ceramics. *Phys Rev B*, **70**.

93. Zhou, T., Zha, J. W., Cui, R. Y., Fan, B. H., Yuan, J. K., and Dang, Z. M. (2011). Improving dielectric properties of $BaTiO_3$/ferroelectric polymer composites by employing surface hydroxylated $BaTiO_3$ nanoparticles. *ACS Appl Mater Interfaces*, **3**:2184–2188.

94. Zhou, Y., Wang, L., Ma, Y. H., Chen, D., Zhao, C. W., and Yang, W. T. (2015). Flexible /PET/BaTiO(3)/ layer-layer composite film with enhanced dielectric properties fabricated by highly loaded /BaTiO(3)/ coating with acrylic resin as binder. *J Appl Polym Sci*, **132**.

95. Zhu, M., Huang, X. Y., Yang, K., Zhai, X., Zhang, J., He, J. L., and Jiang, P. K. (2014). Energy storage in ferroelectric polymer nanocomposites filled with core-shell structured polymer@$BaTiO_3$ nanoparticles: understanding the role of polymer shells in the interfacial regions. *ACS Appl Mater Interfaces*, **6**:19644–19654.

Chapter 4

Graphitic Nanomaterials and Their Polymer Nanocomposites for Dielectric Applications

Bin Li[a] **and Wei-Hong Zhong**[b]
[a]*Department of Mechanical Engineering, Wichita State University, 1845 Fairmount Street, Wichita, Kansas 67260-0133, USA*
[b]*School of Mechanical and Materials Engineering, Washington State University, 405 NE Spokane Street, Pullman, Washington 99164-2920, USA*
bin.li@wichita.edu, katie_zhong@wsu.edu

The large difference in the electrical conductivity between insulating polymers and highly conductive nanomaterials is believed to be responsible for strong interfacial polarization in conductive polymer nanocomposites. Thus, the extremely high dielectric permittivity has been frequently achieved in conductive polymer nanocomposites compared with ceramic-modified polymer nanocomposites [9–24]. Figure 4.1 presents one of the highest dielectric permittivity [26]. In this study, exfoliated graphite nanoplates were incorporated in PVDF matrix. The resulting nanocomposites showed very uniform dispersion of nanoplates, leading to very high dielectric permittivity in the entire frequency range.

Polymer Nanocomposites for Dielectrics
Edited by Wei-Hong Zhong and Bin Li

ISBN 978-981-4745-53-6 (Hardcover), 978-981-4745-54-3 (eBook)
www.panstanford.com

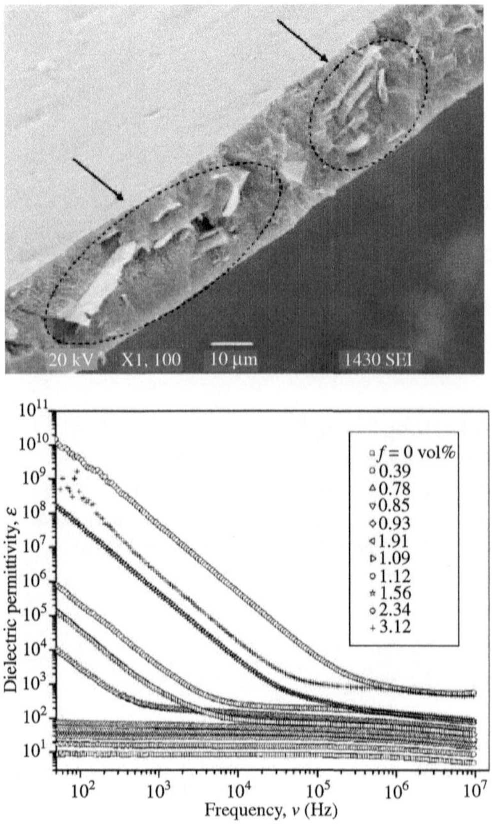

Figure 4.1 SEM image of PVDF nanocomposites with exfoliated graphite nanoplates and their dielectric permittivity. Reprinted from Ref. [26], with permission from John Wiley and Sons.

In contrast to high loadings of ceramic particles in polymer matrix, only a very small amount of conductive nanomaterials is needed in conductive polymer nanocomposites. Typically, significant improvement in dielectric properties can be achieved at a loading level lower than 1.0 vol%. At the same time, the mass density of graphitic carbon materials is around 2 g/cm^3, which is much lower than that of those dielectric ceramics. Therefore, both important characteristics of polymer matrices, the light weight and the flexibility, are well reserved in the resulting nanocomposites. However, it needs to be borne in mind that addition of these highly conductive nanomaterials leads to percolative electrical conduction

[4, 28–41]. This effect can make the resulting nanocomposites unable to withstand high electrical fields. Meanwhile, the electrical conduction contributes to high current leakage. Therefore, for high power capacitor applications, this group of nanocomposites is not as ideal as dielectric ceramic-modified polymer nanocomposites.

4.1 Graphitic Carbon Nanomaterials

Graphitic carbon nanomaterials (GCNs) have been of great interest in multifunctional nanocomposites because of their outstanding mechanical, thermal, and physical properties. Polymer materials modified by GCNs often do not only show enhanced mechanical and thermal properties, but also new functionalities to meet the requirements of diverse applications, such as electromagnetic interference shielding and microwave absorption [42–44]. With the rapid development of nanomanufacturing technology, GCNs with high purity and reduced cost have become increasingly desirable in the fabrication of multifunctional polymer nanocomposites.

Graphene is a one atom thick carbon sheet with a two-dimensional honeycomb lattice consisting of sp^2 hybridized carbons (Fig. 4.2), which can be seen as the fundamental structural unit of graphitic carbon nanomaterials. Graphene is one of the strongest materials with Young's modulus of 1 TPa and tensile strength of 130 GPa, which is 100 times stronger than that of steel. Graphene has unique electronic structures and properties. For example, a single layer of graphene can be considered as a semiconductor at zero distance apart with linear energy dispersion, and the charges that it carries could be considered as massless materials. These massless charges travel at a speed of $\sim 10^6$ ms^{-1} [45]. This unique behavior of graphene has received acknowledgment in the last decade in quantum electrodynamics [25, 45, 46]. In addition, it has been observed that graphene has a theoretical specific surface area close to 2630 m^2/g [25]. This extremely large specific surface area suggests that the interfacial polarization is very significant when graphene nanoparticles are added to a polymer matrix, which contributes to its high dielectric constant. Therefore, the graphene and its oxide-modified polymer nanocomposites have great potential in achieving high dielectric properties [13, 20, 22, 34, 36, 47–62].

Before the successful fabrication of graphene monolayer nanostructures, graphitic carbon nanomaterials were extensively studied [9, 12, 14, 15, 27, 40, 63–74], including graphite nanoplatelets (GNPs) and carbon nanotubes (CNTs). GNPs are composed of a few stacked graphene layers, which are bonded by van der Waals force. The coupling of different layers may diminish the unique electronic structures and properties. The number of layers also affects the physical and chemical properties of GNPs [46]. However, GNPs still possess very high electrical and thermal conductivities and are capable of good mechanical performance. Meanwhile, the fabrication of GNPs is quite facile compared with other graphitic carbon nanomaterials, as they can easily be achieved by exfoliating graphite. This makes GNPs a type of multifunctional and economic modifier for polymer nanocomposites. GNPs, a typical material with a 2D nanostructure, possess a very high surface area that favors interfacial polarization. The extremely high dielectric permittivity presented in Fig. 4.1 was a result of the addition of GNPs in PVDF matrix. By "rolling up" graphene monolayers, we can obtain CNTs. Depending on the number of the layers, there are single-walled CNTs and multiwalled CNTs. CNTs are probably the most well-known 1D nanostructures. The very high aspect ratio and high surface area benefit the electrical and dielectric properties of CNT-modified polymer nanocomposites [7, 44, 70, 75, 76]. Graphene can be "wrapped" to 0D soccer ball–shaped molecules known as fullerene or bucky balls; however, its application in dielectric polymer nanocomposites is not as popular as that of GNPs and CNTs.

4.2 Percolation Phenomenon in Graphitic Nanomaterials/Polymer Nanocomposites

The percolation phenomenon is a characteristic feature of conductive polymer nanocomposites [67]. It is very significant in both electrical and dielectric properties. With increasing loading of conductive nanomaterials, at a critical concentration of nanomaterials, a sharp and abrupt property change can be observed in polymer nanocomposites, for example, the concentration dependence of both dielectric properties and DC conductivity in Fig. 4.3. This is the so-called percolation phenomenon. In terms of dielectric properties, it can be expressed mathematically by Eq. 4.1., where, f_c is the critical concentration, or percolation threshold, f_n is the concentration of nanomaterials, and s is the critical exponent, depending on different polymer nanocomposites.

$$\varepsilon' \propto (f_c - f_n)^s f_c > f_n \tag{4.1}$$

The percolation phenomenon commonly exists in conductive polymer composites, and it is directly related to the network structures of conductive fillers in the insulating polymer matrix. In general, besides the physical properties of these conductive fillers, the formation of this effective network structure, or percolative structure, is strongly dependent on the geometry and the number of conductive fillers. The large number and high aspect ratio help the formation of this network structure. This is why the polymer nanocomposites with conductive nanomaterials, such as the MWCNTs in Fig. 4.3, can possess a high dielectric constant at very low loading levels. At the same loading level, due to the much smaller size of nanomaterials (MWCNTs), the number of MWCNTs is much higher than conductive carbon black (CCB) and short carbon fibers (SCFs). Because MWCNTs, as a material with 1D nanostructure, have very high aspect ratios, we can expect much better dielectric properties in MWCNT/EVA nanocomposites at low loading levels [26]. For CCB, which has a 0D structure (lowest aspect ratio), the concentration of CCB must be very high, in order to achieve similar dielectric properties as EVA/MWCNT nanocomposites.

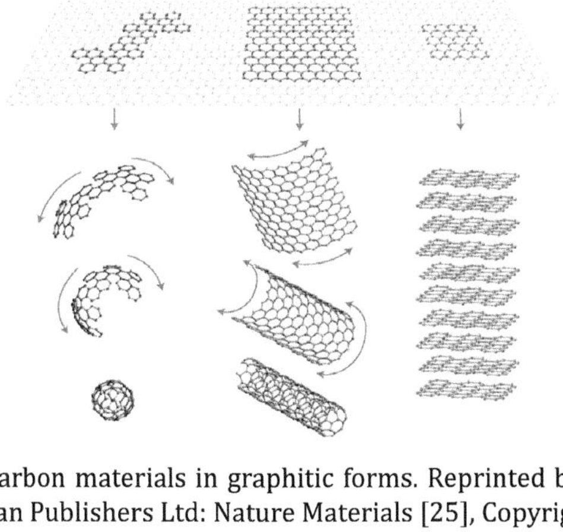

Figure 4.2 Carbon materials in graphitic forms. Reprinted by permission from Macmillan Publishers Ltd: Nature Materials [25], Copyright 2007.

In addition to the geometry and amount of nanomaterials, dispersion and distribution are another two critical factors affecting the percolation phenomenon. In Chapter 1, we have already discussed the significance of the dispersion and distribution of nanomaterials to the properties and functionalities of polymer nanocomposites (Fig. 1.1). The dispersion of nanomaterials is very difficult because their high surface energy tends to aggregate them, which in turn minimizes their surface energy. The dispersion quality of nanomaterials also depends on the aspect ratio of nanomaterials. For example, the 1D nanostructures do not only have high surface energy, but also have strong entanglement because of the high aspect ratio. Both factors trouble the dispersion quality. At last, the dispersion of nanomaterials is also related to processing techniques, as shown in Fig. 4.2. Usually, sonication-assisted solution processing leads to a better dispersion quality, which contributes to the percolation phenomenon at a low loading level of nanomaterials.

4.3 Dispersion and Distribution of Graphitic Nanomaterials

Dispersion of GCNs is critical to the properties and functionalities of polymer/GCN nanocomposites. To achieve high dielectric permittivity at a low loading level, a uniform dispersion is required. The effect of dispersion of GCNs on the dielectric properties of polymer nanocomposites is more complicated than that. In a recent study [27], it was found that the dispersion of carbon nanofibers (CNFs) could induce negative permittivity in polyetherimide (PEI)/CNF nanocomposites, according to Fig. 4.4. When CNFs existed in the PEI matrix mainly in the form of agglomerates consisting of long CNFs (Composite A), the negative permittivity appeared, which usually is only observed in metallic materials, while the improved dispersion resulted in positive permittivity (Composites B and C), and the best dielectric properties were found in the nanocomposites with the highest dispersion quality (Composite C). In this study, it was also found that this negative permittivity was related to the aspect ratio of CNFs.

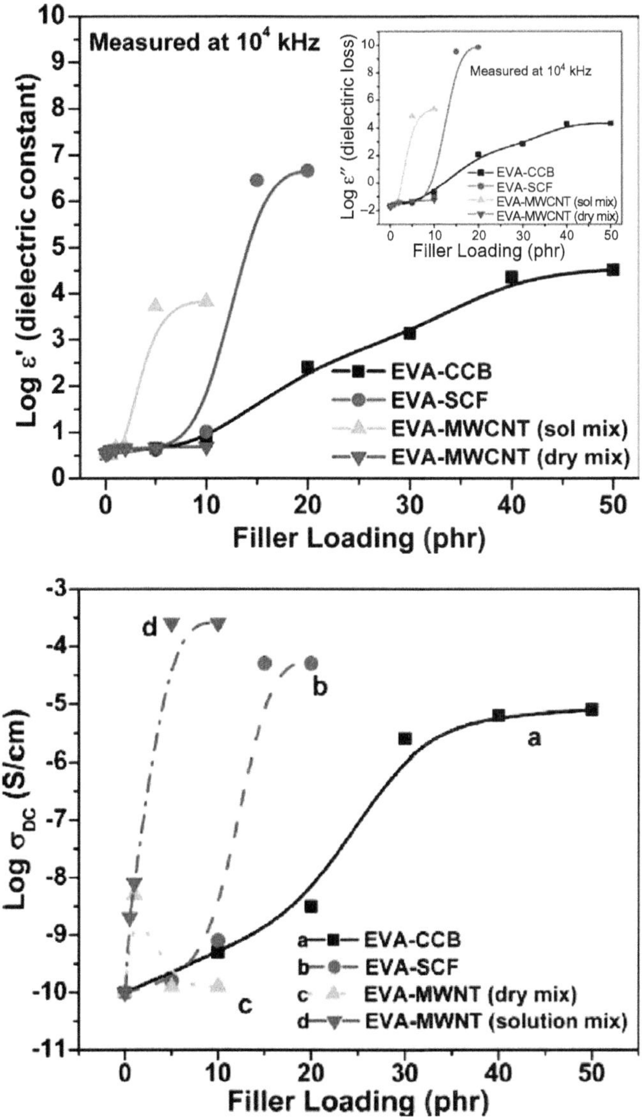

Figure 4.3 Comparison of dielectric properties and DC conductivity of ethylene vinyl acetate composites modified by conductive carbon black (CCB), short carbon fiber (SCF), and multiwalled carbon nanotube (MWCNT). Reprinted from Ref. [26], with permission from John Wiley and Sons.

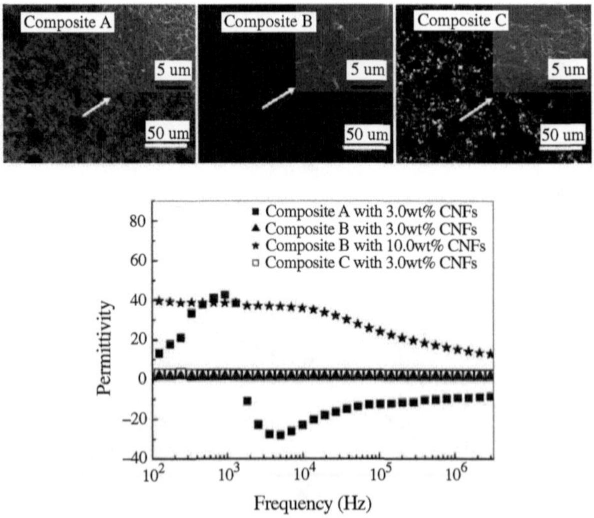

Figure 4.4 Optical microscopy, SEM images, and dielectric permittivity of PEI/CNF nanocomposite films with different dispersion states. Reprinted from Ref. [27], with permission from John Wiley and Sons.

It is clear that, to achieve high dielectric properties, a uniform dispersion of GCNs is required [32]. In the meantime, the orientation and distribution of GCNs in a polymer matrix also matter. Regarding the orientation of GCNs, due to its high electrical conductivity, GCNs can be oriented via the applied electrical field, in addition to mechanical stretching. The orientation of GCNs leads to anisotropic properties of polymer nanocomposites. Compared with anisotropic mechanical and electrical properties, the understanding of their effects on the dielectric properties is still very limited. In the study shown in Figs. 4.5 and 4.6 [4], the single-walled carbon nanotubes (SWCNTs) were aligned in the urethane dimethacrylate–1,6-hexanediol dimethacrylate copolymer (UH) by an AC electrical field with different field strength and frequency. The AC field can induce dipolar interaction among SWCNTs, which contributes to the orientation of SWCNTs. The dielectric properties of the nanocomposites in the direction of the orientation of the SWCNTs were studied. According to Fig. 4.6, the applied field strength had a remarkable impact on the orientation of SWCNTs in the UH matrix and the dielectric constant of the nanocomposites. Only when the field strength was high enough, the orientation of SWCNTs can be activated, contributing to an increase in the dielectric constant. When the electric field strength was above 150 $V_{p\text{-}p}$, the orientation of SWCNTs was saturated, and the dielectric constant did not show any more increase. In addition to the field strength, it was found that the frequency of electrical field also contributes to the orientation of SWCNTs and the dielectric properties of nanocomposites. High frequency favored the orientation of SWCNTs, and thus led to an increased dielectric constant. It should be pointed out that the orientation of SWCNTs, which can be controlled by both field strength and frequency, led to three orders of changes in dielectric properties. This provides an effective way to control the dielectric properties of nanocomposites for different applications.

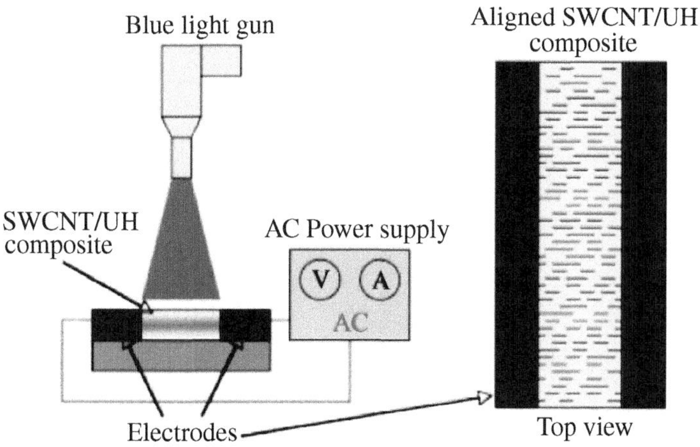

Figure 4.5 Experimental setup of photopolymerization of SWCNT/UH composite under an electric field (left); (b) top view of the alignment cell (right). Reprinted from Ref. [4], with permission from John Wiley and Sons.

Mechanical stretching can also orient nanomaterials, as shown in Fig. 4.7. The degree of orientation of multiwalled carbon nanotubes increased with tensile strain. However, the dielectric properties of the resulting nanocomposites did not show monotonic growth with increasing degree of orientation. Within the range of applied tensile strains, the peak dielectric constant was found at a strain of 12.5%, while between the strain ranges of 25% and 37.5%, the dielectric constant exhibited a monotonic growth (Fig. 4.8). The researcher explained the tensile strain/orientation-dependent dielectric properties by the evolution of a conductive network in the polymer matrix and formation of mini-capacitors.

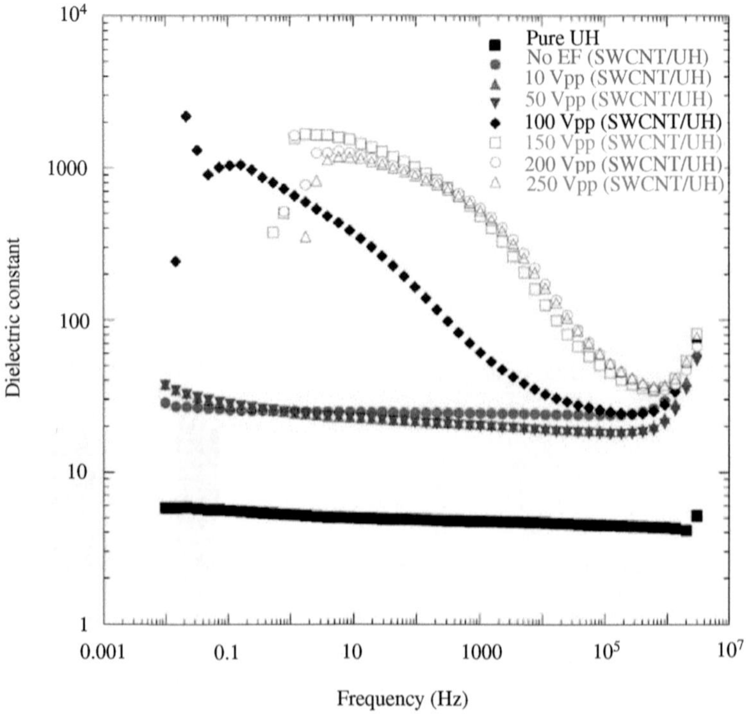

Figure 4.6 Dielectric constants of UH and SWCNTs/UH composites prepared at various electric fields. Reprinted from Ref. [4], with permission from John Wiley and Sons.

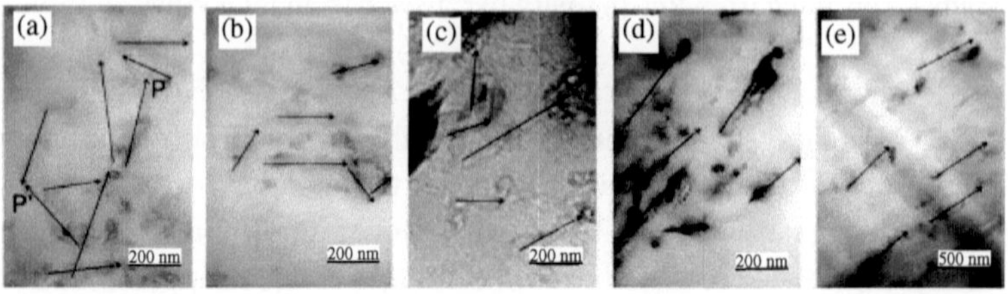

Figure 4.7 TEM images, second-order orientation tensor, and ellipses of samples parallel to the tensile-strain direction in the MWCNT/PVDF nanocomposites with 2.0 wt% MWCNTs respectively: (a) TS = 0; (b) TS = 6.3%; (c) TS = 12.5%; (d) TS = 25.0%; (e) TS = 37.5%. Reprinted with permission from Ref. [6], Copyright 2011, American Chemical Society.

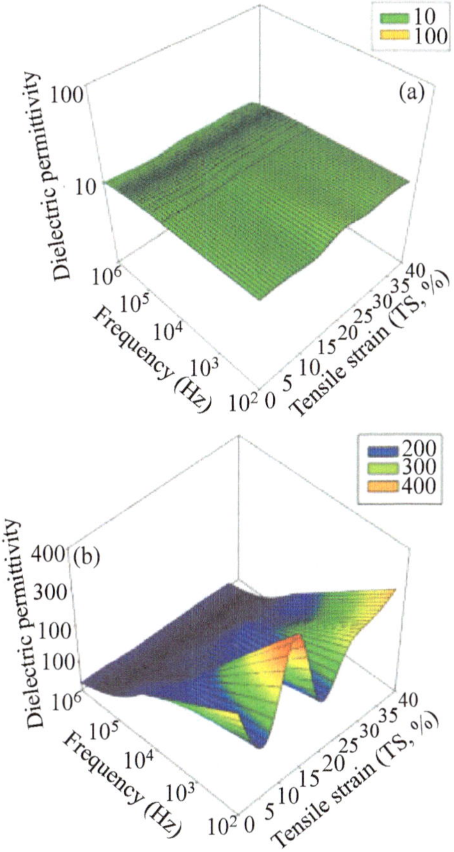

Figure 4.8 (a) Plots of the dielectric permittivity of pure PVDF against the frequency and tensile strain. Dependence of (b) the dielectric permittivity of the MWCNT/PVDF nanocomposites with f_{MWCNT} = 2 wt% on the frequency and tensile strain at room temperature. Reprinted with permission from Ref. [6], Copyright 2011, American Chemical Society.

About these two representative research on the effect of the orientation of carbon nanotubes on the dielectric properties, questions arise. For example, both studies only focused on one single composition. How the concentration of carbon nanotubes affects this orientation-dependent dielectric properties needs to be studied. The orientation of nanomaterials obviously affects the formation of the conductive network, which also relies on the concentration of conductive nanomaterials. So does the aspect ratio of carbon nanotubes, which is critical to the network structure. Meanwhile, both electrical field and mechanical stretching will affect the alignment and conformation of polymer chain structures. For the polymer materials like PVDF, whose crystal structures and dielectric properties are very sensitive to the chain conformation, the synergy between the polymer matrix and nanomaterials may exist. Understanding of this synergy is necessary too.

Distribution of GCNs also plays important roles in the dielectric properties. The best example of the effects of distribution of nanomaterials on the dielectric properties of polymer nanocomposites is the nanocomposites with two incompatible/partially compatible polymers as matrix materials. Because of the differences in surface tension between nanomaterials and polymer materials, nanomaterials typically have a selective distribution in the polymer phases, with which smaller interfacial tension can be achieved. Meanwhile, the viscosity of the polymer matrix should be considered as well in this type of selective distribution. In Fig. 4.9, the incompatible polymer blend of PVDF and low-density polyethylene (LDPE) was used as the matrix for the dielectric polymer nanocomposites. During melting mixing, the MWCNTs selectively distributed in LDPE phase, leading to a double percolation structure (except when PVDF:LDPE = 4:1). Within a certain range of MWCNT concentration, this double percolation structure with a selective MWCNT distribution showed a high dielectric permittivity compared with PVDF/MWCNT nanocomposites and LDPE/MWCNT nanocomposites.

This type of selective distribution does not always make a positive contribution to the dielectric properties of polymer nanocomposites. Figures 4.10 and 4.11 present a study on polymethylmethacrylate (PMMA)/poly(styrene-co-acrylonitrile)(SAN)/MWCNT nanocomposites [1]. PMMA and SAN do not have good compatibility, and therefore, a similar phase separation was observed, as shown in Fig. 4.10. Depending on the ratio of PMMA, both island-like structures and co-continuous phase structures were formed. MWCNTs always prefer to reside in the SAN phase, no matter what the phase structure is. However, this did not lead to any enhanced dielectric permittivity compared with PMMA/MWCNT and SAN/

MWCNT nanocomposites. With the increasing amount of the PMMA phase in the nanocomposites, the dielectric permittivity increases. Combined with the study in Fig. 4.10, we are able to conclude that the effects of selective distribution of carbon nanomaterials also depends on the concentrations of carbon nanomaterials and the properties of the polymer matrices, as well as their interactions with the nanomaterials. These two studies tell us that this selective distribution can be an effective and controllable parameter affecting the dielectric properties of polymer nanocomposites.

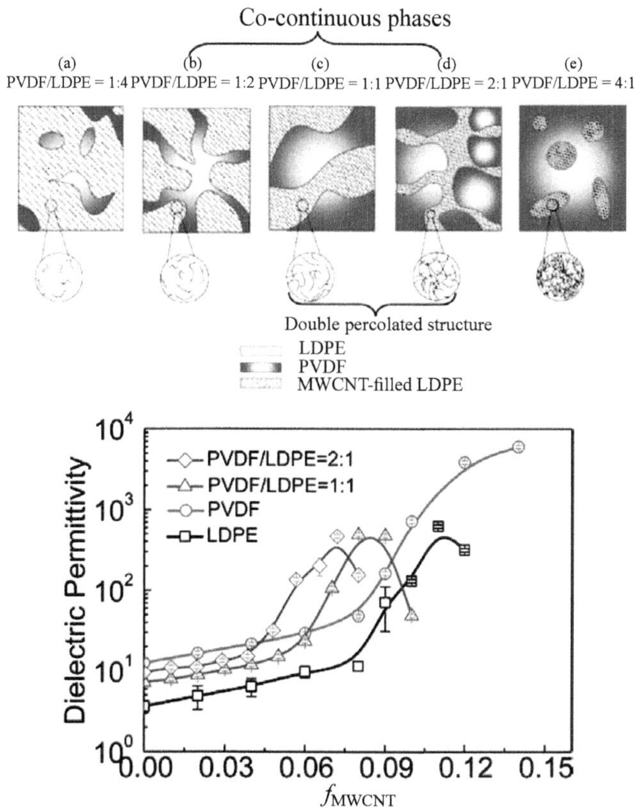

Figure 4.9 (a) Schematic image for the microstructure evolution of the biphasic polymer composites with increasing volume ratio of PVDF/LDPE and (b) dielectric permittivity of biphasic polymer composites as a function of the MWCNT content, measured at 1 Hz and room temperature. Reprinted with permission from Ref. [7], Copyright 2012, American Chemical Society.

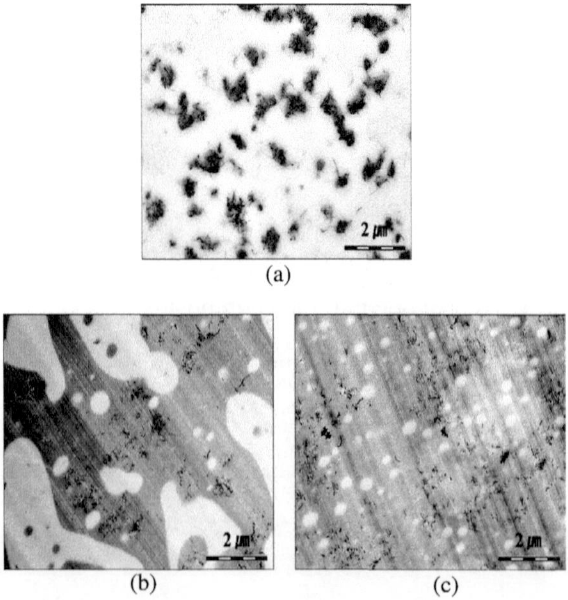

Figure 4.10 TEM images of PMMA/SAN/MWCNT nanocomposites containing 1.5 wt% MWCNT: (a) PMMA/SAN (9/1), (b) PMMA/SAN (5/5), and (c) PMMA/SAN(1/9). Reprinted from Ref. [1], with permission from John Wiley and Sons.

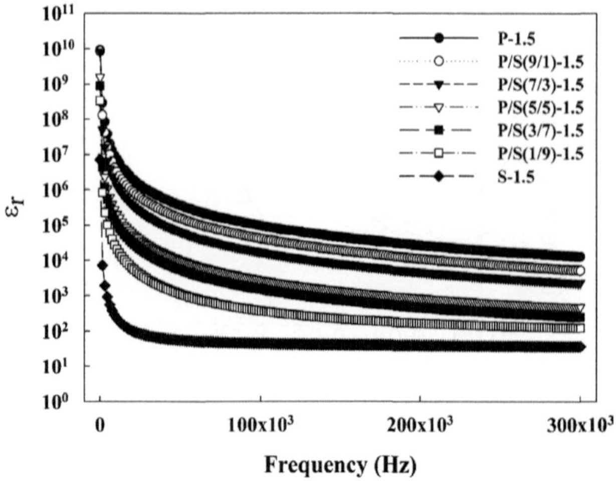

Figure 4.11 Dielectric constants as a function of frequency at various compositions of PMMA and SAN. Reprinted from Ref. [1], with permission from John Wiley and Sons.

4.4 Surface Modification of Nanomaterials

The significant effects of the surface modification of nanomaterials on the properties and functionalities have been addressed in Chapter 3 based on $BaTiO_3$ nanoparticle–reinforced polymer nanocomposites. Surface modification is not only important to the dispersion quality of nanomaterials in the polymer matrix, but also critical to interfacial polarization. Although a great number of studies on the effects of surface modification have been carried out, understanding on this topic is still very limited.

Graphitic carbon nanomaterials have unique surface structures and properties. The uniqueness derives from the sp^2 hybridized carbon structures constraining the π-electron cloud. These surface structures are chemically inert. In order to modify the surface, aggressive surface modification is often necessary to introduce active sites onto the surface of graphitic carbon nanomaterials [77]. Meanwhile, this surface structure also can form some special interactions with certain polymer structures. Raman spectroscopy was applied to study the PVDF/MWCNT nanocomposites prepared by melt mixing procedures (Fig. 4.12) [5]. It was found that there was a molecular-level interaction on the surface of MWCNTs with PVDF chains via the formation of donor–acceptor complexes between delocalized π-electron clouds on MWCNTs and electrophilic fluorine atoms on PVDF. This interaction contributed to the enhanced interfacial polarization and dielectric permittivity. However, this kind of interaction does not exist in most polymer matrices, and therefore, surface modification is still in high demand for the fabrication of high-performance polymer nanocomposites.

The sp^2 hybridized carbons can form π–π interactions with conjugated polymer structures. This type of surface modification is called noncovalent surface modification. In noncovalent surface modification, the surfactant can directly attach to the surface of graphitic carbon nanomaterials without any active sites. Therefore, it is believed that this modification method could maintain the unique structures and properties of graphitic carbon nanomaterials. In the study shown in Fig. 4.12, the emeraldine base (EB) with conjugate structures formed a stable surface coating on MWCNTs through this noncovalent π–π interaction. The resulting nanocomposites with EB-modified MWCNTs did not show much difference in the electrical properties, but the enhancement in dielectric permittivity at low frequencies was observed, indicating a strong interfacial polarization because of this EB surface coating.

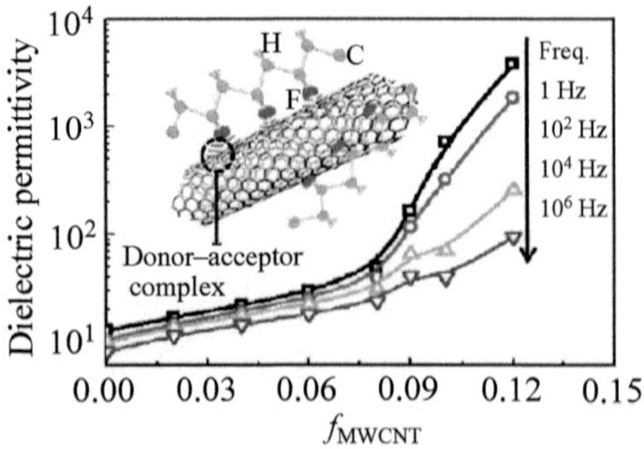

Figure 4.12 Illustration of the polymer wrapping over MWCNTs via the remarkable donor–acceptor complexes and concentration and frequency dependent dielectric permittivity. Reprinted with permission from Ref. [5], Copyright 2011, American Chemical Society.

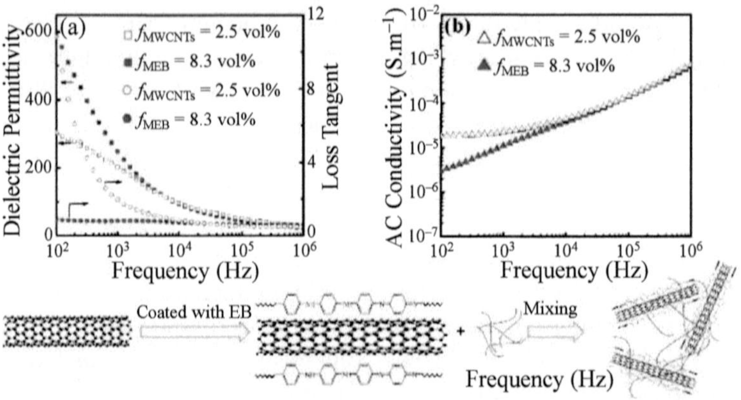

Figure 4.13 Schematic diagrams of the preparation of MEB/PVDF composites and their dielectric and electric properties. Reprinted with permission from Ref. [2], Copyright 2011, American Chemical Society.

Contrary to noncovalent surface modification, traditional covalent surface modification generates a covalent bonding between surfactant molecules and the surface of nanomaterials [23, 67, 68, 70, 75, 78]. As mentioned earlier, active sites on GCNs are needed on the surface to react with surfactant molecules. Typically, various oxidation processes are applied to graphitic carbon nanomaterials to introduce reactive groups, such as carboxylic groups and hydroxyl groups. Although this process damages the sp^2 hybridized structures and deteriorate the functionalities of graphitic nanomaterials, the resulting nanocomposites rarely showed dielectric properties inferior to the nanocomposites with noncovalently modified graphitic nanomaterials.

Silane coupling agents are most commonly used surfactants in covalent surface modification with huge diversity. In a study on poly(lactic acid) (PLA)/graphene oxide (GO) nanocomposites (Fig. 4.14) [8], various silane coupling agents were applied to modify graphene oxide in the presence of ionic liquid. The surface modification contributed to the improvement in dielectric permittivity. At the same time, different molecular structures of silane agents indicate different interfacial polarization mechanisms, and thus different dielectric properties, as shown in Fig. 4.14.

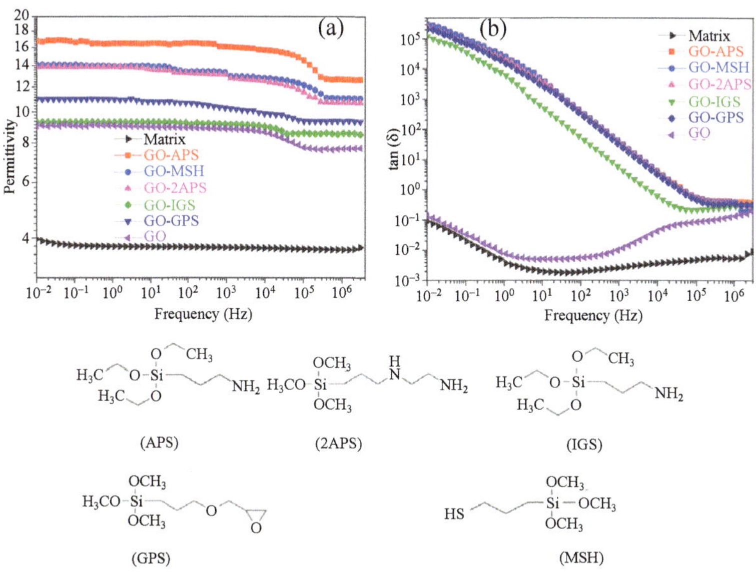

Figure 4.14 Dielectric properties of the PLA nanocomposites with 5 wt% graphene oxide without and with surface modifications by various silane coupling agents: 3-aminopropyl triethoxysilane (APS), 3-isocyanatopropyl triethoxysilane(IGS), 3-glycidyloxypropyl trimethoxysilane (GPS), 3-(2-aminoethylamino) propyl trimethoxylsilane (2APS), and 3-mercaptopropyl trimethoxysilane (MSH). Reprinted with permission from Ref. [8], Copyright 2014, American Chemical Society.

It should be mentioned that, in addition to surface modification of GCN nanoparticles, there are also some other modification processes intended to enhance dielectric properties. For example, the edge structure of GCN particles has high chemical activity. The modification of edges can also contribute to the dispersion of nanomaterials as well as the polarization of nanocomposites [13]. Hybridizing with various dopants, such as metals and ceramics, is also an effective approach to controlling the dielectric properties of polymer nanocomposites, which will be discussed in Chapter 8.

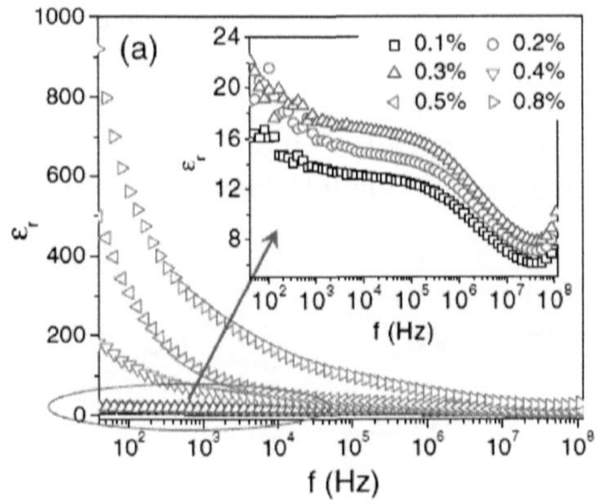

Figure 4.15 Frequency dependency of dielectric constant of solvothermally reduced graphene (SRG)/PVDF composite with various filler contents [3].

Finally, there are some issues regarding GCN/polymer nanocomposites that need to be understood and solved. The dielectric properties of GCN/polymer nanocomposites are very sensitive to their material compositions and structures, such as impurities and dispersion. Because of the limitation of our knowledge, we often found inconsistent research results from different research groups, which also happen to be dielectric studies on other polymer nanocomposites. For example, in Fig. 4.15 [3], the remarkable enhancement in the dielectric constant was observed in PVDF/solvothermally reduced graphene oxide (SRG). However, compared with the research results in Fig. 4.1 [26], the improvement is not sufficient. Considering larger specific surface area, the graphene nanoparticle-modified PVDF nanocomposites should have a stronger interfacial polarization, which is opposite to the results in Figs. 4.1 and 4.15. Possible factors may include the purity, surface properties, geometry, and dispersion of nanomaterials, the processing and post-treatment of polymer nanocomposites, and so on. The excellent dielectric permittivity of PVDF/pristine GNP nanoparticles in Fig. 4.1 does not only indicate the great potential of GCN nanomaterials in dielectric polymer nanocomposite applications, but also suggests that there is more to be explored in the dielectric properties of polymer nanocomposites.

References

1. Lee, M., et al. (2011). Morphology and electrical properties of polymethylmethacrylate/poly(styrene-co-acrylonitrile)/multi-walled carbon nanotube nanocomposites. *J Appl Polym Sci*, **121**(2):743–749.
2. Zhou, T., et al. (2011). Surface-functionalized MWNTs with emeraldine base: preparation and improving dielectric properties of polymer nanocomposites. *ACS Appl Mater Interfaces*, **3**(12):4557–4560.
3. He, L. X. and S. C. Tjong (2013). Low percolation threshold of graphene/polymer composites prepared by solvothermal reduction of graphene oxide in the polymer solution. *Nanoscale Res Lett*, **8**.
4. Park, C., et al. (2006). Aligned single-wall carbon nanotube polymer composites using an electric field. *J Polym Sci, Part B: Polym Phys*, **44**(12):1751–1762.
5. Yuan, J. K., et al. (2011). Giant dielectric permittivity nanocomposites: realizing true potential of pristine carbon nanotubes in poly(vinylidene fluoride) matrix through an enhanced interfacial interaction. *J Phys Chem C*, **115**(13):5515–5521.
6. Yao, S. H., et al. (2011). Stretch-modulated carbon nanotube alignment in ferroelectric polymer composites: characterization of the orientation state and its influence on the dielectric properties. *J Phys Chem C*, **115**(40):20011–20017.
7. Yuan, J. K., et al. (2012). Biphasic polymer blends containing carbon nanotubes: heterogeneous nanotube distribution and its influence on the dielectric properties. *J Phys Chem C*, **116**(2):2051–2058.
8. Fu, Y., L. S. Liu, and J. W. Zhang (2014). Manipulating dispersion and distribution of graphene in PLA through novel interface engineering for improved conductive properties. *ACS Appl Mater Interfaces*, **6**(16):14069–14075.
9. Abu-Abdeen, M., A. S. Ayesh, and A. Aljaafari (2011). Dielectric relaxation and rheological properties of single-walled carbon

nanotubes reinforced poly(3-octylthiophene-2,5-diyl). *J Thermoplast Compos Mater,* **26**(5):605–626.

10. Agarwal, M., et al. (2009). Conductive paper from lignocellulose wood microfibers coated with a nanocomposite of carbon nanotubes and conductive polymers. *Nanotechnology,* **20**(21).
11. Al-Ghamdi, A. A., et al. (2009). Stability of new electrostatic discharge protection and electromagnetic wave shielding effectiveness from poly(vinyl chloride)/graphite/nickel nanoconducting composites. *Polym Degrad Stab,* **94**(6):980–986.
12. Al-Hartomy, O. A., et al. (2011). Influence of graphite nanosheets on the structure and properties of PVC-based nanocomposites. *J Appl Polym Sci,* **120**(6):3628–3634.
13. Almadhoun, M. N., et al. (2014). Influence of stacking morphology and edge nitrogen doping on the dielectric performance of graphene-polymer nanocomposites. *Chem Mater,* **26**(9):2856–2861.
14. Al-Saleh, M. H., et al. (2013). Impedance characteristics and conductivity of CNT/ABS nanocomposites. *J Phys D: Appl Phys,* **46**(38).
15. Al-Saleh, M. H., S. A. Jawad, and H. M. Ghanem (2014). Electrical and dielectric behaviors of dry-mixed CNT/UHMWPE nanocomposites. *High Perform Polym,* **26**(2):205–211.
16. Ameli, A., et al. (2014). Polypropylene/carbon nanotube nano/microcellular structures with high dielectric permittivity, low dielectric loss, and low percolation threshold. *Carbon,* **71**:206–217.
17. Arjmand, M., et al. (2013). An innovative method to reduce the energy loss of conductive filler/polymer composites for charge storage applications. *Compos Sci Technol,* **78**:24–29.
18. Ayesh, A. S., S. S. Ibrahim, and A. A. Aljaafari (2012). Electrical, optical, and rheological properties of ozone-treated multiwalled carbon nanotubes-polystyrene nanocomposites. *J Reinf Plast Compos,* **32**(6):359–370.
19. Baseghi, S., et al. (2015). Lightweight high-density polyethylene/carbonaceous nanosheets microcellular foams with improved electrical conductivity and mechanical properties. *J Mater Sci,* **50**(14):4994–5004.
20. Bhattacharya, P. and C. K. Das (2012). Investigation on microwave absorption capacity of nanocomposites based on metal oxides and graphene. *J Mater Sci: Mater Electron,* **24**(6):1927–1936.
21. Chang, K. C., et al. (2014). Enhancement of physical properties of electroactive polyimide nanocomposites by addition of graphene nanosheets. *Polym Int,* **63**(6):1011–1017.

22. Chen, Y., et al. (2013). Preparation of thermostable PBO/graphene nanocomposites with high dielectric constant. *Nanotechnology*, **24**(24):245702.

23. Chen, Z., S. Chen, and J. Zhang (2011). Effect of functionalized MWCNTs with surfactant and coupling agent on properties of LDPE/POE blends. *J Elastomers Plast*, **43**(6):543–558.

24. Dang, Z. M., et al. (2008). Supersensitive linear piezoresistive property in carbon nanotubes/silicone rubber nanocomposites. *J Appl Phys*, **104**(2).

25. Geim, A. K., and K. S. Novoselov (2007). The rise of graphene. *Nature Materials*, **6**(3): 183–191.

26. He, F., et al. (2009). High dielectric permittivity and low percolation threshold in nanocomposites based on poly(vinylidene fluoride) and exfoliated graphite nanoplates. *Adv Mater*, **21**(6):710–715.

27. Li, B. and W. H. Zhong (2010). Influence of carbon nanofiber network variability on the AC conductivity of polyetherimide composite films. *Macromol Mater Eng*, **295**(4):310–314.

28. Li, B. and W. H. Zhong (2011). Review on polymer/graphite nanoplatelet nanocomposites. *J Mater Sci*, **46**(17):5595–5614.

29. Linares, A., et al. (2014). Broadband dielectric spectroscopy of nanocomposites based on PVDF and expanded graphite. 2nd International Conference on Structural Nano Composites (Nanostruc 2014), **64**.

30. Liu, F., et al. (2014). Crystalline properties, dielectric response and thermal stability of in-situ reduced graphene oxide/poly(vinylidene fluoride) nanocomposites. *IEEE Trans Dielectr Electr Insul*, **21**(4):1446–1454.

31. Long, Y. and X. B. Liu (2014). Dielectric, mechanical and thermal properties of novel core-shell CuPc@MWCNTs/PEN composite films. *J Mater Sci: Mater Electron*, **25**(2):1089–1096.

32. Lovell, C. S., et al. (2009). Thermodynamic approach to enhanced dispersion and physical properties in a carbon nanotube/polypeptide nanocomposite. *Polymer*, **50**(8):1925–1932.

33. Maxwell, K. S., et al. (2010). Finite element analysis of a three-phase piezoelectric nanocomposite. *J Intell Mater Syst Struct*, **21**(11):1073–1084.

34. Mensah, B., et al. (2014). A study of graphene oxide-reinforced rubber nanocomposite. *J Appl Polym Sci*, **131**(16).

35. Mitra, S., A. Singha, and D. Chakravorty (2013). Non-linear temperature variation of resistivity in graphene/silicate glass nanocomposite. *J Phys D: Appl Phys*, **46**(37).

36. Monti, M., et al. (2013). Morphology and electrical properties of graphene-epoxy nanocomposites obtained by different solvent assisted processing methods. *Compos Part A: Appl Sci Manuf*, **46**:166–172.

37. Moudam, O., et al. (2013). Carbon nanotube-epoxy nanocomposites: correlation and integration of dynamic impedance, dielectric, and mechanical analyses. *J Nanomater*, 801850.

38. Nayak, L., D. Khastgir, and T. K. Chaki (2013). A mechanistic study on electromagnetic shielding effectiveness of polysulfone/carbon nanofibers nanocomposites. *J Mater Sci*, **48**(4):1492–1502.

39. Ohlan, A., et al. (2010). Microwave absorption behavior of core-shell structured poly(3,4-ethylenedioxy thiophene)-barium ferrite nanocomposites. *ACS Appl Mater Interfaces*, **2**(3):927–933.

40. Olowojoba, G., et al. (2013). Influence of process parameters on the morphology, rheological and dielectric properties of three-roll-milled multiwalled carbon nanotube/epoxy suspensions. *Polymer*, **54**(1):188–198.

41. Ounaies, Z., et al. (2008). Evidence of piezoelectricity in SWNT-polyimide and SWNT-PZT-polyimide composites. *J Thermoplast Compos Mater*, **21**(5):393–409.

42. Wang, Y., et al. (2013). Preparation of graphene/$BaFe_{12}O_{19}$-Ni0.8Zn0.2Fe_2O_4 nanocomposite and its microwave absorbing properties. *J Sol-Gel Sci Technol*, **67**(2):344–350.

43. Yang, Y., et al. (2013). Electrospun magnetic carbon composite fibers: synthesis and electromagnetic wave absorption characteristics. *J Appl Polym Sci*, **127**(6):4288–4295.

44. Basuli, U., et al. (2012). Electrical properties and electromagnetic interference shielding effectiveness of multiwalled carbon nanotubes-reinforced EMA nanocomposites. *Polym Compos*, **33**(6):897–903.

45. Seo, J.-W. T., et al. (2011). High-concentration aqueous dispersions of graphene using nonionic, biocompatible block copolymers. *J Phys Chem Lett*, **2**(9):1004–1008.

46. Mahmood, N., et al. (2014). Graphene-based nanocomposites for energy storage and conversion in lithium batteries, supercapacitors and fuel cells. *J Mater Chem A*, **2**(1):15–32.

47. An, N. L., et al. (2015). Preparation and properties of beta-phase graphene oxide/PVDF composite films. *J Appl Polym Sci*, **132**(10).

48. Ansari, S., and E. P. Giannelis (2009). Functionalized graphene sheet-poly(vinylidene fluoride) conductive nanocomposites. *J Polym Sci, Part B: Polym Phys*, **47**(9):888–897.

49. Ataur Rahman, M., et al. (2013). Fabrication and characterization of highly efficient flexible energy harvesters using PVDF–graphene nanocomposites. *Smart Mater Struct*, **22**(8):085017.

50. Bernholc, J., et al. (2008). Multiscale simulations of high performance capacitors and nanoelectronic devices. *Proceedings of the HPCMP Users Group Conference*, 2008:221–228.

51. Cho, S., J. S. Lee, and J. Jang (2015). Poly(vinylidene fluoride)/NH_2-treated graphene nanodot/reduced graphene oxide nanocomposites with enhanced dielectric performance for ultrahigh energy density capacitor. *ACS Appl Mater Interfaces*, **7**(18):9668–9681.

52. Greco, A., A. Timo, and A. Maffezzoli (2012). Development and characterization of amorphous thermoplastic matrix graphene nanocomposites. *Materials*, **5**(10):1972–1985.

53. Hong, J. Y., E. Lee, and J. Jang (2013). Electro-responsive and dielectric characteristics of graphene sheets decorated with TiO_2 nanorods. *J Mater Chem A*, **1**(1):117–121.

54. Zhang, X. J., et al. (2014). Enhanced microwave absorption property of reduced graphene oxide (RGO)-$MnFe_2O_4$ nanocomposites and poly (vinylidene fluoride). *ACS Appl Mater Interfaces*, **6**(10):7471–7478.

55. Wu, Y., et al. (2015). Exceptional dielectric properties of chlorine-doped graphene oxide/poly (vinylidene fluoride) nanocomposites. *Carbon*, **89**:102–112.

56. Zhang, W. L. and H. J. Choi (2013). Fabrication and electrorheology of graphene oxide/ionic N-substituted copolyaniline composite. *Colloid Polym Sci*, **291**(6):1401–1408.

57. Romasanta, L. J., et al. (2011). Functionalised graphene sheets as effective high dielectric constant fillers. *Nanoscale Res Lett*, **6**.

58. Wissert, R., et al. (2010). Graphene nanocomposites prepared from blends of polymer latex with chemically reduced graphite oxide dispersions. *Macromol Mater Eng*, **295**(12):1107–1115.

59. He, L. T. and Sie Chin (2013). Low percolation threshold of graphene/polymer composites prepared by solvothermal method of graphene oxide in the polymer solution. *Nanoscale Res Lett*, **8**.

60. Sharmila, T. K. B., et al. (2014). Microwave exfoliated reduced graphene oxide epoxy nanocomposites for high performance applications. *Polymer*, **55**(16):3614–3627.

61. Tong, W. S., et al. (2014). Novel method for the fabrication of flexible film with oriented arrays of graphene in poly(vinylidene fluoride-co-hexafluoropropylene) with low dielectric loss. *J Phys Chem C*, **118**(20):10567–10573.

62. Thomassin, J. M., et al. (2014). Poly(methyl methacrylate)/graphene oxide nanocomposites by a precipitation polymerization process and their dielectric and rheological characterization. *Macromolecules*, **47**(6):2149–2155.

63. Yan, W., et al. (2013). Designing atmospheric-pressure plasma sources for surface engineering of nanomaterials. *Plasma Chem Plasma Process*, **33**(2):479–490.

64. Ursache, S., et al. (2013). Dielectric and electromagnetic behavior of conductive nanocomposites polymers: PP/MWCNT investigations for EMI applications. *Interdiscip Res Eng: Steps Towards Breakthrough Innovation Sustainable Dev*, **8–9**:353–360.

65. Sui, G., et al. (2008). Dielectric properties and conductivity of carbon nanofiber/semi-crystalline polymer composites. *Acta Mater*, **56**(10):2381–2388.

66. Zhang, Y. H., et al. (2012). Dielectric properties of carbon-nanotube/amino-functionalized poly(arylene ether ketone) composites. *High Perform Polym*, **24**(3):173–179.

67. Wang, F., et al. (2009). Dielectric properties of epoxy composites with modified multiwalled carbon nanotubes. *Polym Bull*, **63**(1):101–110.

68. Hayashida, K. (2013). Dielectric properties of polymethacrylate-grafted carbon nanotube composites. *RSC Adv*, **3**(1):221–227.

69. Zheng, P. L., et al. (2014). Effect of multiwalled carbon nanotubes on the crystallization and dielectric properties of BP-PEN nanocomposites. *J Mater Sci: Mater Electron*, **25**(9):3833–3839.

70. Kar, P. and A. Choudhury (2013). Electrical and dielectric properties of polyaniline doped with carboxyl-functionalized multiwalled carbon nanotube. *Adv Polym Technol*, **32**:E760–E770.

71. Yu, C. R., et al. (2011). Electrical and dielectric properties of polypropylene nanocomposites based on carbon nanotubes and barium titanate nanoparticles. *Compos Sci Technol*, **71**(15):1706–1712.

72. Tjong, S. C., G. D. Liang, and S. P. Bao (2007). Electrical behavior of polypropylene/multiwalled carbon nanotube nanocomposites with low percolation threshold. *Scr Mater*, **57**(6):461–464.

73. Zheng, W. and S. C. Wong (2003). Electrical conductivity and dielectric properties of PMMA/expanded graphite composites. *Compos Sci Technol*, **63**(2):225–235.

74. Li, Q., et al. (2008). Large dielectric constant of the chemically functionalized carbon nanotube/polymer composites. *Compos Sci Technol*, **68**(10–11):2290–2296.

75. He, G. W., et al. (2011). Preparation, morphology and properties of acylchloride-grafted multiwall carbon nanotubes/fluorinated polyimide composites. *Compos Sci Technol*, **71**(16):1914–1920.

76. Potschke, P., et al. (2004). Rheological and dielectrical characterization of melt mixed polycarbonate-multiwalled carbon nanotube composites. *Polymer*, **45**(26):8863–8870.

77. Fragiadakis, D. and P. Pissis (2007). Glass transition and segmental dynamics in poly(dimethylsiloxane)/silica nanocomposites studied by various techniques. *J Non-Cryst Solids*, **353**(47–51):4344–4352.

78. Gunasekaran, S. G., et al. (2013). Siloxane core dianhydride modified ether linked cyclohexyl diamine based multi-walled carbon nanotube reinforced polyimide (MWCNT/PI) nanocomposites. *J Polym Res*, **21**(1).

Chapter 5

Poly(Vinylidene Fluoride) in Dielectric Nanocomposites

Bin Li[a] **and Wei-Hong Zhong**[b]
[a]*Department of Mechanical Engineering, Wichita State University, 1845 Fairmount Street, Wichita, Kansas 67260-0133, USA*
[b]*School of Mechanical and Materials Engineering, Washington State University, 405 NE Spokane Street, Pullman, Washington 99164-2920, USA*
bin.li@wichita.edu, katie_zhong@wsu.edu

The commercial plastic dielectric films today are dominated by those made of polypropylene and polyester because of their high electrical breakdown strength, low dielectric loss, self-healing ability, low cost, etc. However, their low energy density has been the biggest challenge for their applications in the future generation of capacitors.

Fluoropolymers, on the other hand, are considered as a family of the most promising polymer materials for future dielectric devices. In particular, the semicrystalline poly(vinylidene fluoride) (PVDF) shows high dielectric polarization leading to high energy density because of its highly polar molecular structures and unique polymorphism of crystal structures. Meanwhile, besides the high energy density, PVDF possesses distinct piezoelectricity and

Polymer Nanocomposites for Dielectrics
Edited by Wei-Hong Zhong and Bin Li

ISBN 978-981-4745-53-6 (Hardcover), 978-981-4745-54-3 (eBook)
www.panstanford.com

pyroelectricity, suggesting its great potential in energy harvesting, sensor, and actuator applications. Therefore, PVDF has gained great attention in academia. It has been frequently studied as the matrix materials for polymer nanocomposites for various dielectric applications. In the previous chapters, the dielectric properties of many PVDF nanocomposites containing dielectric ceramic nanoparticles and graphitic carbon nanomaterials were introduced. In contrast to polypropylene and polyester, PVDF has high dielectric polarization that leads to its high dielectric loss, which is the primary barrier to its application in dielectric capacitors for energy storage. The relatively low breakdown voltage and high leakage current are critical issues to PVDF as well. To achieve high dielectric properties, and increase breakdown voltage and resistance to current leaking, both synthesis of copolymers of PVDF [14, 25, 26, 29, 36, 37, 40, 42–45, 48, 55, 60–62, 68, 69, 71, 72] and fabrication of their nanocomposites have been widely studied [1–13, 30, 59, 70].

Dielectric properties of PVDF are strongly dependent on their crystal structures; both copolymer structures and incorporating nanomaterials can remarkably affect the crystal structures of PVDF and its copolymers. In other words, in PVDF nanocomposites, besides the dispersion of nanomaterials and interfacial interaction between nanomaterials and the polymer matrix, as discussed in Chapters 3 and 4, the crystal structures of PVDF are a critical factor for the dielectric properties, too. Therefore, the review of dielectric properties of fluoropolymer nanocomposites in this chapter will primarily address how nanomodifications affect the crystal structures of polymer matrices and their impact on the dielectric performances of the resulting nanocomposites, while the effects of copolymer structures will be discussed in Chapter 6.

5.1 Crystal Structures of PVDF

PVDF is a semicrystalline polymer with a highly polar chemical structure ($-CH_2-CF_2-$), providing strong permanent dipole moment. However, it is the crystal structures that control the dielectric polarization in PVDF. The flexible chains of PVDF are susceptible to different processing conditions, such as temperature, mechanical stretching, and electrical field-poling, showing different chain conformations, which determine different crystal structures, as shown in Fig. 5.1

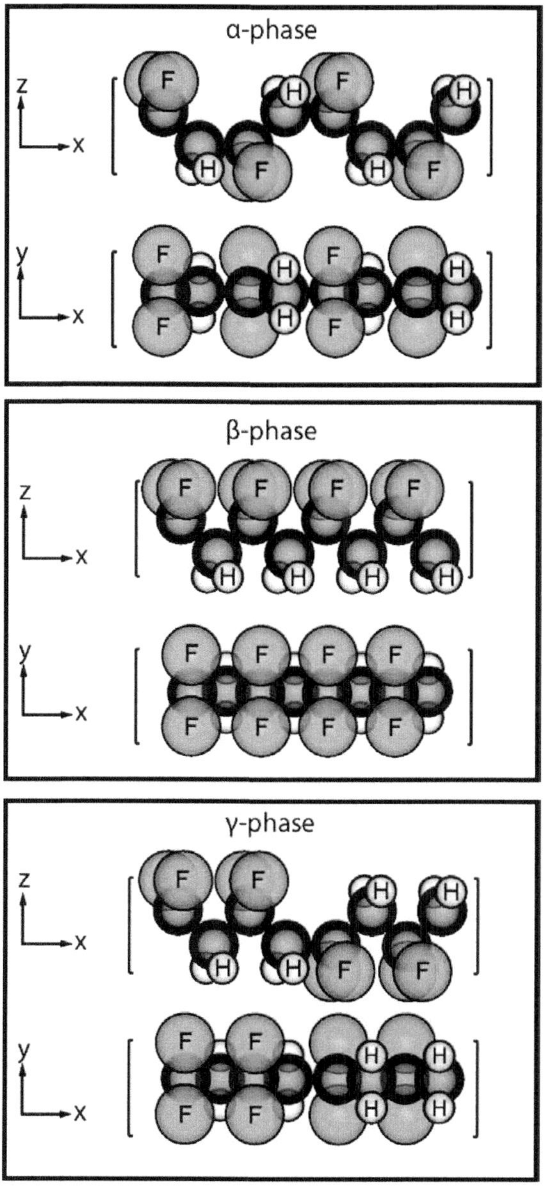

Figure 5.1 Schematic illustration of polymorphism of PVDF crystal structures. Reprinted with permission from Ref. [35], Copyright 2015, American Chemical Society.

A few crystal structures have been found in PVDF structures, designated as α-phase, β-phase, γ-phase, and δ-phase, while the first three phases are more commonly observed in PVDF. α-phase is the most common and also the most stable PVDF crystal structure, showing a TGTG′ conformation (T: trans; G: guache). In this conformation, the neighboring dipole moments are in the opposite directions, and thus, α-phase is a nonpolar phase showing zero net dipole moment [33]. In β-phase, the PVDF chains have an all-trans conformation (TTTT), which leads to the highest spontaneous polarization compared with that of other PVDF structures. β-phase is also responsible for the piezoelectricity and pyroelectricity in PVDF. Therefore, β-phase PVDF has been of the greatest interests for PVDF-based dielectric materials for various applications. Usually, β-phase can be obtained via hot stretching and electrically poling α-phase PVDF (when the electrical field is above 400 MV/m [26]) [20, 27, 38]. γ-phase is another highly polar PVDF structure, which can be seen as the intermediate phase between α- and β-phases, showing a TTTG–TTTG′ conformation. γ-phase has received limited attention compared with α- and β-phases. Usually, γ-phase can form under high temperature, high pressure, or by adding certain inorganic salts to PVDF during crystallization of PVDF [46, 47]. But achieving high fraction of γ-phase in PVDF remains a challenge.

It is noteworthy to mention that hot-stretching is a convenient and popular approach to achieving high volume fraction of β-phase in PVDF [15, 65–67]. For PVDF nanocomposites, hot-stretching has a more complex impact on the structures and dielectric properties of the nanocomposites. The hot-stretching process could induce phase transformation to highly polar β-phase. In the meantime, hot-stretching also changes the dispersion and distribution of nanomaterials in the polymer matrix, as shown in Fig. 5.2.

In Chapters 3 and 4, we have already discussed the significance of both dispersion and distribution of nanomaterials in properties and functionalities of the polymer nanocomposites. The changes in the dispersion and distribution of nanomaterials can certainly affect the percolative network of functional nanomaterials, which is critical to the electrical and dielectric properties of polymer nanocomposites. Figure 5.2 presents an example of this complex situation [75]. In this study on PVDF/Ni nanocomposites, different stretching ratios were applied. The resulting nanocomposites showed remarkable enhancement in both electrical conductivity and dielectric constant [75]. The X-ray diffraction study revealed that the increase in the stretching ratio led to a higher fraction of β-phase in the nanocomposites. According to the optical images in Fig. 5.2, it is clear that the dispersion and distribution of Ni nanoparticles were completely changed after stretching. At last, previous studies [66] found that nanomaterials often induce the formation of β-phase PVDF. Therefore, this effect should be considered as well in order to obtain an in-depth understanding of the relationship between the structures of PVDF nanocomposites and their dielectric properties.

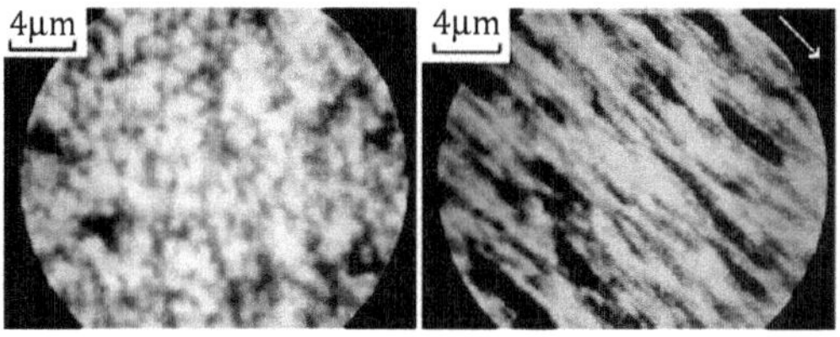

Figure 5.2 Dispersion and distribution of Ni nanoparticles in PVDF/Ni nanocomposites before and after hot-stretching. Reprinted from Ref. [75], with permission from John Wiley and Sons.

5.2 Effects of Processing on PVDF Crystal Structures

PVDF is a thermoplastic with good melt processability [21, 22] and solubility in various organic solvents. Both melt processing (compression molding) and solution casting methods have been frequently used to fabricate PVDF dielectric films and their nanocomposites. The processing parameters have complicated effects on the crystal structures of PVDF. Typically, the solution casting creates more β-phase in PVDF, in particular, when dimethylformamide (DMF) or dimethyl sulfoxide (DMSO) [31] was used as the solvent, while compression molding tends to generate a high fraction of the more stable α-phase. By hot-stretching α-phase PVDF, the phase transformation will take place. Typically, in order to achieve a high fraction of β-phase, the stretching temperature is controlled around 80 to 120°C, while the stretching ratio is between

4 and 5 times [17, 18]. Lower stretching temperature favors a higher content of β-phase. However, the inverse phase transformation could be easily induced by annealing PVDF at an elevated temperature, usually above 120°C for a few hours. At higher temperatures, the less stable β-phase will gradually recrystallize and transform to the more stable α-phase. For PVDF films made from both melt processing and solution casting, they usually possess both α-phase and β-phase, and their portions can be tuned via the aforementioned hot-stretching and annealing processes.

Table 5.1 summarizes the effects of different processing for making PVDF films on crystal structures. On the basis of this table, we can see that the fabrication methods and stretching temperature have significant impacts on the crystal structures of PVDF films, including the crystallinity and melting temperature as well as fractions of α- and β-phases, respectively. In addition, the crystal structures in PVDF are usually mixtures of both α- and β-phases, no matter what processing procedures are applied to the PVDF films.

5.3 Effects of Nanomaterials on Phase Transformation in PVDF

The role of nanomaterials as a heterogeneous nucleation agent for semicrystalline polymers has been well established. Meanwhile, the unique surface properties of different nanomaterials with and without surface modification frequently contribute to the phase transformation in semicrystalline polymers. However, the complexity of both nanomaterials and semicrystalline polymer materials and different processing approaches often led to inconsistent research findings in similar polymer nanocomposites. Therefore, although numerous studies on this topic have been reported, the understanding of the effects of nanomaterials on crystal structures of PVDF and crystal structure dependent dielectric properties is still in an early stage. At the same time, in some studies on dielectric properties of PVDF nanocomposites, the effects of nanomaterials on crystal structures of PVDF were not even discussed [24, 49, 54, 73, 74], although they provided excellent insights into other important factors such as particle size, interface, and dispersion.

Table 5.1 Effects of processing on crystal structures of PVDF films

Films	Processing method	$T_{stretch}$ (°C)	Crystallinity (wt %)[a]	T_m (°C)[b]	α (wt %)[b]	β (wt %)[b]	$D_{(110)\alpha}$ (nm)[b]	$D_{(110/200)\beta}$ (nm)[b]
A	biax. stretched		49.4	176	27.5	19.6	37	25
B	hot-pressed, uniax. stretched	90	32.2	143	18.2	30.6	14	11
C	hot-pressed, uniax. stretched	130	35.6	133	32.9	7.3	19	16
D	solution-cast, uniax. stretched							
⊥[c]		60	43.3	146	14.1	0	23	
||[d]		60	43.3	146	21.1	2.7	11	10
E	solution-cast, uniax. stretched							
⊥[c]		110	41.1	146	9.6	0	25	
||[d]		110	41.1	146	22.3	9.1	13	12

Source: Ref. [34]

[a]Determined by differential scanning calorimetry (DSC). [b]Determined by 2D wide-angle X-ray diffraction (WAXD). [c]PVDF crystalline lamellae lie flat-on in the film with their *c*-axes parallel to the film normal. [d]PVDF crystalline lamellae lie transverse in the film with their *c*-axes parallel to the drawing direction.

In Chapter 2, various effective medium theories by simplifying MWS equations have been introduced for evaluation of dielectric properties of polymer nanocomposites. In these equations (Eqs. 2.18–2.20), only the volume fractions and dielectric properties of both nanomaterials and a polymer matrix were considered. Therefore, according to these equations, it is reasonable to only discuss the effects of particle size and dispersion state of nanomaterials on the dielectric properties of polymer nanocomposites, considering the large difference between the dielectric properties of dielectric nanomaterials and polymer materials. However, according to the original MWS equation (Eq. 2.14) derived from a double-layer model for the description of the interfacial polarization, it should be pointed out that, in Eqs. 2.18–2.20, the same layer spacing was assumed in both layers, which was often not the reality in polymer nanocomposite. Meanwhile, the relaxation time of this double-layer model was related to the conductivities of both polymer materials and nanomaterials. It has been reported that the phase transformation in PVDF could lead to the changes in AC conductivity and conduction mechanism [17, 66]. Therefore, the effects of the crystal structures on the interfacial polarization might be more significant and complicated.

In a study on PVDF/ $BaTiO_3$ nanocomposites [23], there was no phase transformation observed with the addition of up to 30 wt% $BaTiO_3$ nanoparticles to PVDF matrix. Both pure PVDF and the nanocomposites exhibited α-phase only. Meanwhile, the nanoparticles led to a decrease in crystallinity, which was believed to be a result of the decreased infusion heat of PVDF with an increase in $BaTiO_3$ content. It is speculated that the processing plays a very important role in this finding. In this study, the solid paste of PVDF and $BaTiO_3$ powder was prepared by mixing acetone as a thinner. The solid paste was molded into the nanocomposite samples for characterizations. The processing of nanocomposites may affect both dispersion/distribution of nanomaterials and crystallization of polymers.

The studies conducted by Javier González-Benito [56, 64] also revealed the effects of $BaTiO_3$ nanoparticles on crystal structure PVDF in PVDF/ $BaTiO_3$ nanocomposites fabricated by a different processing technique. In these two studies, the PVDF powder and $BaTiO_3$ powder were mixed together via a high energy cryo-ball milling method, and the nanocomposite films were made by hot-pressing the ball milled powder mixture. The crystal structures of PVDF in the nanocomposites in both powder form and film form were analyzed. In the form of powder, the fraction of β-phase increased as a result of ball milling processing, and the addition of $BaTiO_3$ nanoparticles enhanced this effect of ball milling and increased crystallinity of the PVDF matrix. The compression molding recovered the α-phase. In addition to the phase transformation, the observation of the nanocomposite films prepared by compression molding also indicated the dependence of morphologies of PVDF crystals on $BaTiO_3$ content, as shown in Fig. 5.3. The aspect ratio of the crystal lamellae increased with the $BaTiO_3$ content. By ultrasound-assisted dispersion of $BaTiO_3$ in DMF and solution mixing, the PVDF/$BaTiO_3$ nanocomposite films were fabricated by casting the nanocomposite solutions [52, 53].

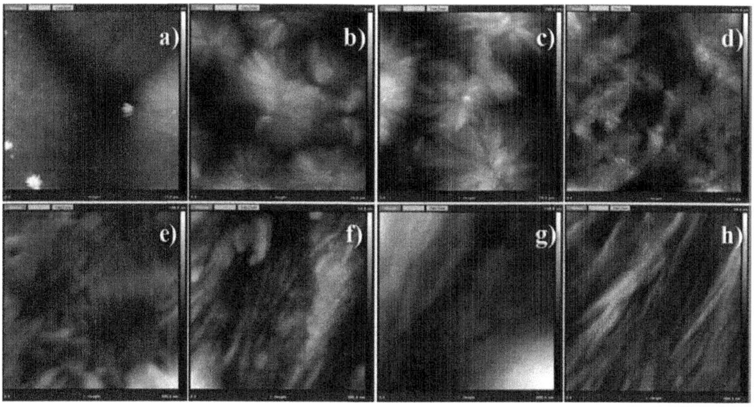

Figure 5.3 Effects of $BaTiO_3$ content on the crystal morphologies of PVDF: (a and e) 0%; (b and f) 1%; (c and g) 5%; and (d and h) 10%. Top images 20 mm × 20 mm and bottom images 800 nm × 800 nm. Reprinted from Ref. [56], with permission from John Wiley and Sons.

The studies on these nanocomposite films concluded that it was the particle size, not the $BaTiO_3$ content, which contributed to the phase transformation to β-phase, because of the size-dependent interactions between nanoparticles and polymer matrix. Smaller particles tended to act as nucleating agents, while large particles with reduced interfacial area often led to the formation of structural defects and decreased the crystallinity.

Another example of this complex topic is the study of effects of interfacial interactions between PVDF matrix and the nanoparticles in the PVDF/$BaTiO_3$ nanocomposites. It was found that not only the pristine BT, but also the surface modification, had limited effects on the crystal structures of PVDF. According to a comprehensive analysis by XRD, DSC, and FTIR, no obvious phase transformation was observed [28]. The resulting enhancement in energy storage was largely due to the addition of $BaTiO_3$ and the interfacial effects as a result of surface modification of $BaTiO_3$ via a titanate couple agent, as shown in Fig. 3.10 in Chapter 3.

Obviously, in the face of these inconsistent results in similar nanocomposite systems, and considering the different fabrication techniques applied, a comprehensive and systematic study is needed to examine the contributions of each factor, including the structures and morphologies of both nanomaterials and polymer matrix, as well as processing parameters. In particular, it is highly possible that the crystal structures of PVDF, including phase structure, morphologies, and crystallinity, actually depend on the synergetic effects of these factors. In other words, the nucleation or anti-nucleation effects of the nanoparticles might only occur under certain processing conditions, such as critical temperature and solvents. However, research on such synergetic effects is still in its infancy, in spite of the tremendous amount of findings reported on the crystal structures of PVDF and other semicrystalline polymers.

It seems that graphitic nanomaterials favor the phase transformation of PVDF compared with $BaTiO_3$ nanoparticles [7, 19, 66, 67]. In a study of PVDF/CNF nanocomposites [3], both XRD and DSC analysis revealed the phase transformation of PVDF with addition of CNFs, as shown in Figs. 5.4 and 5.5. According to Fig. 5.4, the addition of CNFs to PVDF matrix can induce β-phase crystal structures, even without stretching the nanocomposite films. XRD results in Fig. 5.4 also revealed the enhanced crystallization to α-phase as evidenced by stronger diffraction peaks at $2\theta = 18.7°$ and 27.3°, suggesting that CNF is a good nucleating agent for both α- and β-phases. The isothermal recrystallization in this study was conducted by reheating the nanocomposite film to 200°C for 2 hours and then was kept at 165°C for 5 hours. As expected, the isothermal recrystallization could lead to an increase in the α-phase of PVDF. At the same time, we can still observe strong diffraction peaks of β-phase. This fact further supports the role of CNFs as a nucleation agent for the β-crystal. The understanding of the formation mechanism is still not clear. Several mechanisms have been proposed, such as the zigzag carbon structures on CNFs and ultrasonication cavitation.

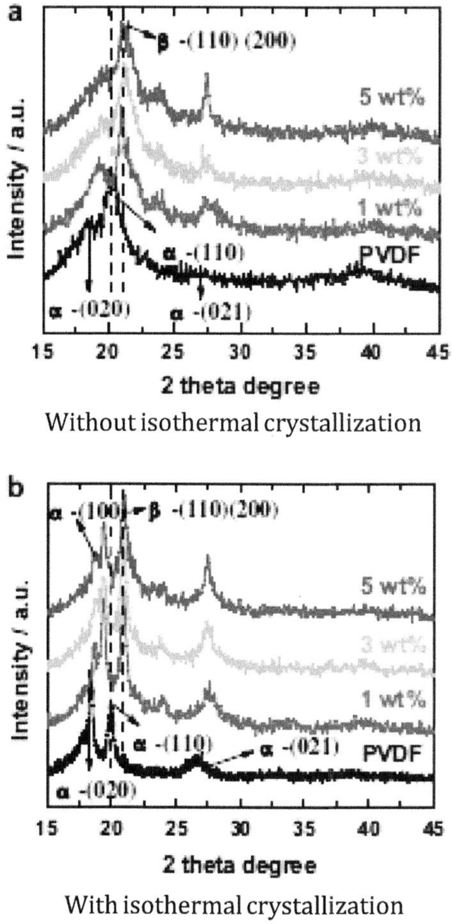

Figure 5.4 XRD spectra of PVDF/CNF nanocomposites fabricated via solution casting method. Reprinted from Ref. [66], Copyright 2010, with permission from Elsevier.

Meanwhile, in the same study, DSC results further proved the findings in XRD spectra (Fig. 5.5). The broadening of the melting peak around 161°C in the recrystallized sample suggests the effects of CNFs on nucleating β-crystals. The melting temperatures of α- and β-crystals are very close, which is probably the reason why there was only one melting peak in Fig. 5.5a. The isothermal recrystallization seems to have some positive effects on distinguishing these two crystal phases, according to the blunting (two peaks) of the melting peak around 161°C. For the nanocomposites containing 3 and 5 wt%, a melting peak at 177°C was observed, which represents the γ-phase formed directly from the melt at 161°C [66]. In another research on PVDF/CNF nanocomposites [17], mechanical stretching was applied to the nanocomposite films to induce β-crystals. It was also found that at the same stretching ratio, the increase of CNF content also led to higher fractions of β-phase. But this research did not discuss if the CNF alone could induce β-phase. Multiwalled carbon nanotubes (MWCNTs) have also shown positive results in nucleating β-phase in PVDF nanocomposites [63] with some exceptions [39].

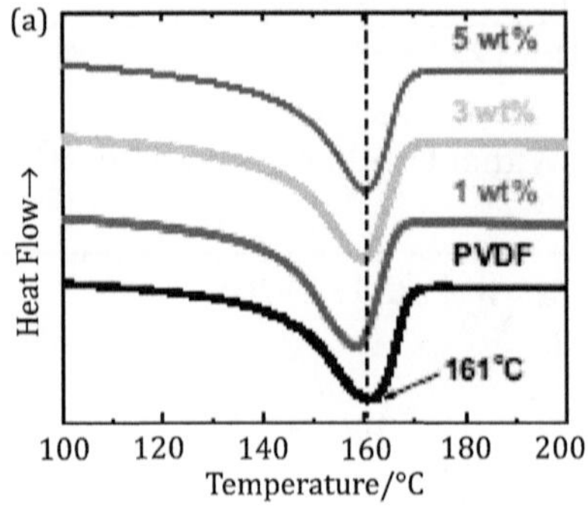

Without isothermal crystallization

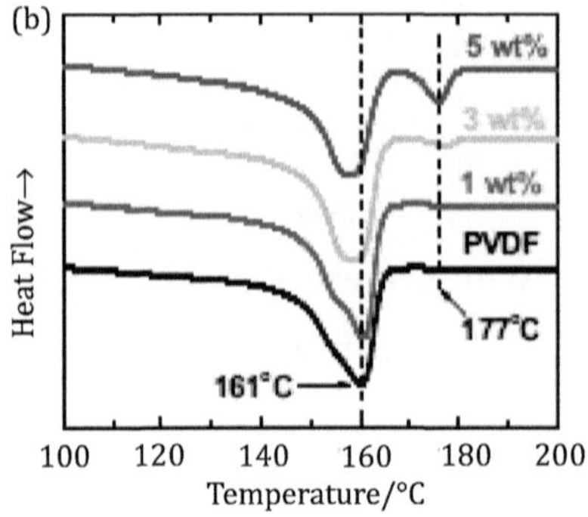

With isothermal crystallization

Figure 5.5 DSC curves of PVDF/CNF nanocomposites with and without isothermal recrystallization. Reprinted from Ref. [66], Copyright 2010, with permission from Elsevier.

5.4 Effects of Surface Properties of Nanomaterials on Phase Transformation in PVDF

The surface properties of nanomaterials are critical to interface-related processes and phenomena, including nucleation of crystal structures of PVDF on the surface of nanomaterials. Although the understanding of the pristine surface of nanomaterials on the nucleation of β-phase PVDF is still unclear, many attempts have been made to explore how the surface properties of nanomaterials affect the formation of β-phase. The most direct path to achieving this goal is surface modification on nanomaterials. Figure 5.6 presents an example of the surface modification of reduced graphene oxide via NH_2-treated graphene nanodots. To understand how the surface properties affect the fraction of β-phase in PVDF, this research compared different nanoparticles with and without surface modifications, as listed in Table 5.2. The comprehensive analysis was conducted by FTIR, XRD, and DSC. First of all, it is obvious that both graphene nanodots (GND) and reduced graphene oxide (RGO) could contribute to a higher fraction of β-phase compared with pure PVDF; secondly, the existence of $-NH_2$ on GND surface can benefit the increase of fraction of β-phase; at last, the surface modification of RGO by NH_2-GND resulted in the highest fraction of β-phase. This systematic study directly revealed the effects of surface properties on the formation of β-crystals. Meanwhile, it also provides an alternative to create a high fraction of β-crystals, in addition to the mechanical stretching of PVDF nanocomposites.

The addition of MWCNTs promoted the formation of β-phase in PVDF, but decreased the intensity of the diffraction peaks of α-phase [58]. The surface coating by ZrO_2 also favored the formation of β-phase. It was proposed that the better interfacial interactions between PVDF and MWCNT would lead to all-trans conformation of PVDF chains on the surface of the coated MWCNT and, therefore, promote the formation of β-phase. This phase transformation as a result of the surface coating of ZrO_2 contributed to an increase in the dielectric constant within the entire testing range, as shown in Fig. 5.7. Furthermore, this study compared different coating methods: isothermal hydrolysis of $ZrOCl_2$, and chemical precipitation by the reaction between $ZrOCl_2 \cdot 8H_2O$ and NH_4OH. These two different coating approaches led to a remarkable difference in the dielectric properties of the resulting nanocomposites. The study did not address the reasons for this difference. It is speculated that in addition to the chemical structures, other factors such as crystal structures and the defects of the coating also affected the interactions with the PVDF matrix and MWCNTs, leading to different crystal structures and dielectric properties of the resulting nanocomposites.

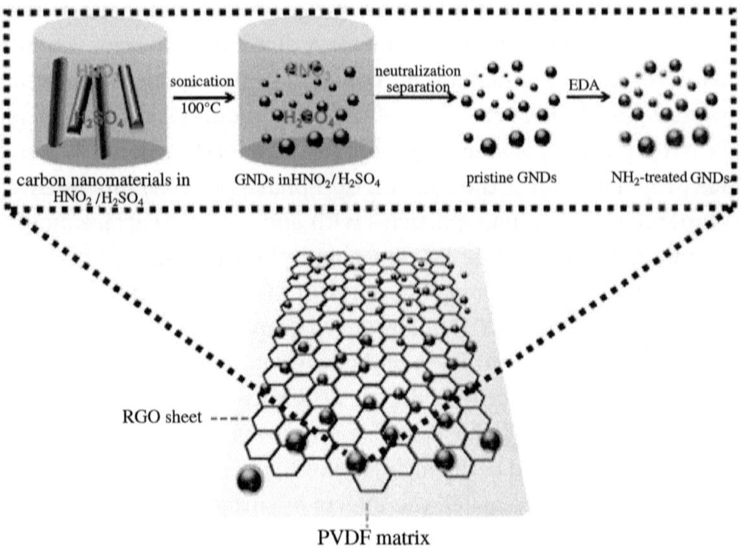

Figure 5.6 Surface modification of reduced graphene oxide (RGO) via NH_2-treated graphene nanodots (GND). Reprinted with permission from Ref. [16], Copyright 2015, American Chemical Society.

According to the research in Fig. 5.7, if the interactions between the surfaces of nanomaterials and PVDF could favor the all-trans arrangement of PVDF chains, the resulting nanocomposites can show a higher fraction of β-crystals. In particular, if the surface structures tend to have strong attractive interactions with C–F bonds or C–H bonds in PVDF, the possibility of achieving all-trans conformation seems to be higher. One effective approach is to use nanoparticles with intrinsic electrostatic charges or to create surface charges on particle surfaces [32, 50, 51].

Table 5.2 Relative proportion of β-phase in the pristine PVDF and nanocomposites

Sample[a]	FT-IR[b]	XRD[c]	DSC[d]
Pristine PVDF	0.29		0.19
PVDF/GND	0.44	0.51	0.29
PVDF/RGO	0.59	0.53	0.35
PVDF/NH_2-GND	0.69	0.57	0.58
PVDF/NH_2-GND/RGO	0.78	0.64	0.68

Source: Ref. [16]
[a]The concentration of all fillers is 10 vol% with respect to PVDF nanocomposites. [b]The value was calculated using Eq. 5.1. [c]The value was calculated using Eq. 5.2. [d]The value was estimated from the multiple melting peaks.

$$\frac{I_\beta}{I_\alpha + I_\beta} = \frac{I_{513} + I_{839} + I_{1272}}{I_{491} + I_{530} + I_{615} + I_{764} + I_{796} + I_{978} + I_{513} + I_{839} + I_{1272}} \tag{5.1}$$

where the I_α and I_β denote the peak intensity for α- and β-phases, respectively. I_{491}, I_{513}, I_{530}, I_{615}, I_{764}, I_{796}, I_{839}, I_{978}, and I_{1272} indicate the peak intensity of FTIR absorption bands at 491, 515, 530, 615, 764, 796, 839, 978, and 1272 cm^{-1}, respectively.

$$\frac{I_\beta}{I_\alpha + I_\beta} = \frac{I_{200/110}}{(I_{100} + I_{020} + I_{110} + I_{021})_\alpha + (I_{200/110})_\beta} \tag{5.2}$$

where the I_{100}, I_{020}, I_{110}, I_{021}, and $I_{200/110}$ mean the XRD peak intensity at (100), (020), (110), (021), and (200/110) planes, respectively.

In the study of the effects of the surface properties of nanomaterials on PVDF crystal structures, it was found that electrostatic charges of the nanoparticles played a critical role in nucleating the β-crystals of PVDF [50]. In this study, $CoFe_2O_4$ nanoparticles with negative electrostatic charges were employed to modify the crystal structures of PVDF. In order to verify the effects of surface charges in forming β-phase, three different surfactants were used: anionic surfactant (SDS), nearly neutral surfactant (Trioton), and cationic surfactant (CTAB). According to zeta potential test, the pristine $CoFe_2O_4$ nanoparticles had the highest negative surface charges, and the resulting nanocomposite had the highest fraction of β-crystals (90%) characterized by FTIR, while the SDS-treated nanoparticles modified PVDF were second to it with 30% β-crystal. In the nanocomposites containing the other two nanoparticles, there was no β-crystal found on FITR spectra. The mechanism was proposed that the negative surface charges of the nanoparticles could form strong attraction to the $-CH_2$ groups with a high positive charge density, leading to an all-trans transformation, and eventually enhanced piezoelectric coefficient, as shown in Fig. 5.8. However, the research did not provide satisfactory explanation on why there were no useful interactions between positively charged cationic surfactant and $-CF_2$ groups with high negative charge density to lead to the similar all-trans chain conformation.

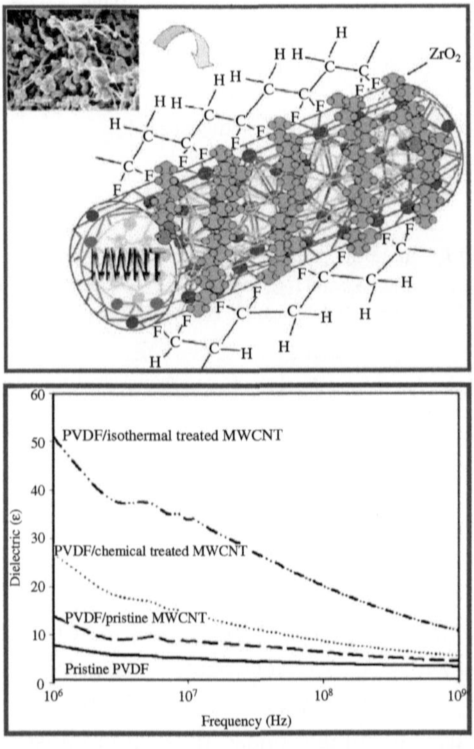

Figure 5.7 Schematic diagram showing the assumed mechanism for the formation of β-phase crystal for PVDF/ZrO_2-coated MWCNT composites and their dielectric properties. Reprinted with permission from Ref. [58], Copyright 2010, American Chemical Society.

Obviously, there is more to be explored in this approach. Meanwhile, it is necessary to evaluate the effects of the surface charges on other dielectric properties, in addition to dielectric constant, such as break strength and leakage current, if the resulting nanocomposites are intended for energy storage applications, since the surface charges also contribute to the electrical conduction in these nanocomposites.

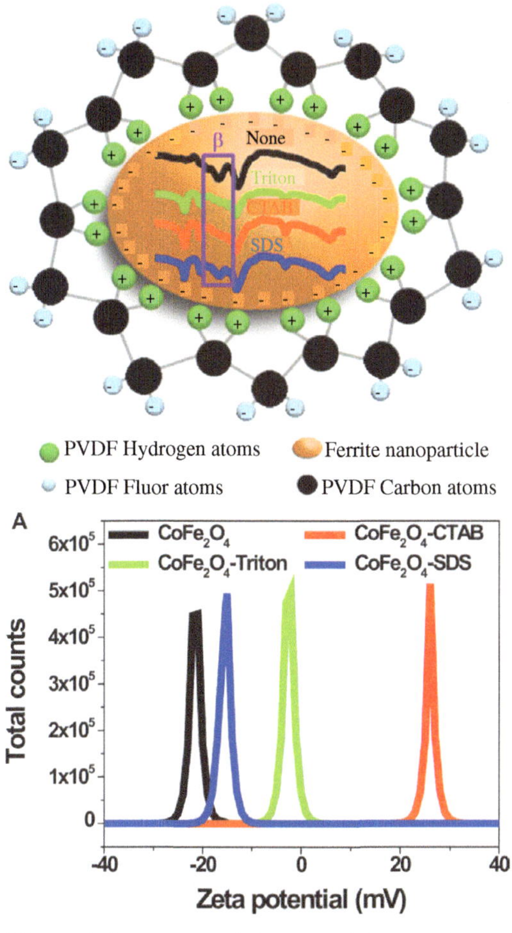

Figure 5.8 Schematic illustration of electrostatic interactions between $-CH_2$ in PVDF and surfaces of $CoFe_2O_4$ nanoparticles (upper) and zeta potential of various $CoFe_2O_4$ without and with different surfactants (lower). Reprinted with permission from Ref. [50], Copyright 2012, American Chemical Society.

In the end, it should be pointed out again that the popularity of the β-phase of PVDF is based on its highest polarity and strong piezo- and pyroelectric signals. PVDF with high fractions of β-phase is an ideal electroactive polymer material for various sensor and actuator applications. Therefore, a large portion of PVDF research in this area has been achieving the same goal: to increase the content of β-phase. However, we also need to face the disadvantages of β-phase. In particular, for energy storage applications, the high loss of β-phase has been one of the biggest obstacles for PVDF in dielectric capacitor applications that are dominated by BOPP and PET films today. Therefore, the high energy density of β-PVDF is not gaining practical application in dielectric capacitors. Therefore, the studies on the low loss α-phase PVDF has been gaining more and more attention today [41, 57, 76]. Meanwhile, the stability of β-phase is always a critical issue in practical applications. It appears that the existence of nanomaterials in PVDF has little effect on this stability issue, which is also insufficiently studied. To improve its stability, the copolymer structures are more effective. In Chapter 6, the use of copolymers and their nanocomposites will be introduced.

References

1. Abdelaziz, M. (2013). Investigations on optical and dielectric properties of PVDF/PMMA blend doped with mixed samarium and nickel chlorides. *J Mater Sci: Mater Electron*, **24**, 2727–2736.
2. Abdelaziz, M. and Abdelrazek, E. M. (2014). The effect of nanographite addition on the physical properties of poly(vinylidene fluoride)/hydroxypropyl cellulose blend. *J Mater Sci:Mater Electron*, **25**, 5481–5490.
3. Adireddy, S., Puli, V. S., Lou, T. J., Elupula, R., Sklare, S. C., Riggs, B. C., and Chrisey, D. B. (2015). Polymer-ceramic nanocomposites for high energy density applications. *J Sol-Gel Sci Technol*, **73**, 641–646.
4. Aepuru, R. and Panda, H. S. (2014). Adsorption of charge carriers on radial zinc oxide and the study of their stability and dielectric behavior in poly(vinylidene fluoride). *J Phys Chem C*, **118**, 18868–18877.
5. Aftab, S., Hall, D. A., Aleem, M. A., and Siddiq, M. (2013). Low field ac study of PZT/PVDF nano composites. *J Mater Sci: Mater Electron*, **24**, 979–986.
6. Alluri, N. R., Saravanakumar, B., and Kim, S. J. (2015). Flexible, hybrid piezoelectric film ($BaTi(1\text{-}x)ZrxO_3$)/PVDF nanogenerator as a self-powered fluid velocity sensor. *ACS Appl Mater Interfaces*, **7**, 9831–9840.
7. An, N. L., Liu, S. L., Fang, C. Q., Yu, R. E., Zhou, X., and Cheng, Y. L. (2015). Preparation and properties of beta-phase graphene oxide/PVDF composite films. *J Appl Polym Sci*, **132**.

8. Andrews, C., Lin, Y. R., Tang, H. X., and Sodano, H. A. (2011). Influence of aspect ratio on effective electromechanical coupling of nanocomposites with lead zirconate titanate nanowire inclusion. *J Intell Mater Syst Struct*, **22**, 1879–1886.
9. Ansari, S. and Giannelis, E. P. (2009). Functionalized graphene sheet-poly(vinylidene fluoride) conductive nanocomposites. *J Polym Sci Part B: Polym Phys*, **47**, 888–897.
10. Ataur Rahman, M., Lee, B.-C., Phan, D.-T., and Chung, G.-S. (2013). Fabrication and characterization of highly efficient flexible energy harvesters using PVDF–graphene nanocomposites. *Smart Mater Struct*, **22**, 085017.
11. Baji, A., Mai, Y. W., Li, Q., and Liu, Y. (2011). Nanoscale investigation of ferroelectric properties in electrospun barium titanate/poly (vinylidene fluoride) composite fibers using piezoresponse force microscopy. *Compos Sci Technol*, **71**, 1435–1440.
12. Baji, A., Mai, Y. W., Abtahi, M., Wong, S. C., Liu, Y., and Li, Q. (2013). Microstructure development in electrospun carbon nanotube reinforced poly(vinylidene fluoride) fibers and its influence on tensile strength and dielectric permittivity. *Compos Sci Technol*, **88**, 1–8.
13. Bauer, F. (2012). Review on the properties of the ferrorelaxor polymers and some new recent developments. *Appl Phys A: Mater Sci Process*, **107**, 567–573.
14. Chan, H. L. W., Zhang, Q. Q., Ng, W. Y., and Choy, C. L. (2000). Dielectric permittivity of PCLT/PVDF-TRFE nanocomposites. *IEEE Trans Dielectr Electr Insul*, **7**, 204–207.
15. Chinaglia, D. L., Gregorio, R., and Vollet, D. R. (2012). Structural modifications in stretch-induced crystallization in PVDF films as measured by small-angle X-ray scattering. *J Appl Polym Sci*, **125**, 527–535.
16. Cho, S., Lee, J. S., and Jang, J. (2015). Poly(vinylidene fluoride)/NH2-treated graphene nanodot/reduced graphene oxide nanocomposites with enhanced dielectric performance for ultrahigh energy density capacitor. *ACS Appl Mater Interfaces*, **7**, 9668–9681.
17. Costa, P., Silva, J., Sencadas, V., Costa, C. M., van Hattum, F. W. J., Rocha, J. G., and Lanceros-Mendez, S. (2009). The effect of fibre concentration on the alpha to beta-phase transformation, degree of crystallinity and electrical properties of vapour grown carbon nanofibre/poly(vinylidene fluoride) composites. *Carbon*, **47**, 2590–2599.
18. da Silva, A. B., Wisniewski, C., Esteves, J. V. A., and Gregorio, R. (2010). Effect of drawing on the dielectric properties and polarization of pressed solution cast beta-PVDF films. *J Mater Sci*, **45**, 4206–4215.

19. Dang, Z. M., Wang, L., Yin, Y., Zhang, Q., and Lei, Q. Q. (2007). Giant dielectric permittivities in functionalized carbon-nanotube/electroactive-polymer nanocomposites. *Adv Mater*, **19**, 852.

20. Darestani, M. T., Chilcott, T. C., and Coster, H. G. L. (2014). Effect of poling time on filtration properties of PVDF membranes treated in intense electric fields. *Polym Bull*, **71**, 951–964.

21. Dmitriev, I. Y., Gladchenko, S. V., Lavrent'ev, V. K., Praslova, O. E., and Elyashevich, G. K. (2006). Dependence of the dielectric constant on the structure of extruded poly(vinylidene fluoride) films. *Russ J Appl Chem*, **79**, 642–646.

22. Dong, L. J., Li, R., Xiong, C. X., and Quan, H. Y. (2010). Effect of heat treatment on the electrical properties of lead zirconate titanate/poly (vinylidene fluoride) composites. *Polym Int*, **59**, 756–758.

23. Elzayat, M. Y. F., El-Sayed, S., Osman, H. M., and Amin, M. (2012). X-ray diffraction and differential scanning calorimetry studies of a $BaTiO_3$/poly(vinylidene fluoride) composites. *Polym Eng Sci*, **52**, 1945–1950.

24. Fan, B. H., Zha, J. W., Wang, D. R., Zhao, J., and Dang, Z. M. (2012). Size-dependent low-frequency dielectric properties in the $BaTiO_3$/poly(vinylidene fluoride) nanocomposite films. *Appl Phys Lett*, **100**.

25. Feng, Y. F., Gong, H. H., Xie, Y. C., Wei, X. Y., Yang, L. J., and Zhang, Z. C. (2015). Strong induced polarity between poly(vinylidene fluoride-co-chlorotrifluoroethylene) and alpha-SiC and its influence on dielectric permittivity and loss of their composites. *J Appl Phys*, **117**.

26. Gadinski, M. R., Han, K., Li, Q., Zhang, G. Z., Reainthippayasakul, W., and Wang, Q. (2014). High energy density and breakdown strength from beta and gamma phases in poly(vinylidene fluoride-co-bromotrifluoroethylene) copolymers. *ACS Appl Mater Interfaces*, **6**, 18981–18988.

27. Gan, W. C. and Abd Majid, W. H. (2014). Effect of TiO_2 on enhanced pyroelectric activity of PVDF composite. *Smart Mater Struct*, **23**.

28. Gao, L., He, J. L., Hu, J., and Li, Y. (2014). Large enhancement in polarization response and energy storage properties of poly(vinylidene fluoride) by improving the interface effect in nanocomposites. *J Phys Chem C*, **118**, 831–838.

29. Gao, L., He, J. L., Hu, J., and Li, Y. (2014). Tailored ferroelectric responses and enhanced energy density in PVDF-based homopolymer/terpolymer blends. *J Appl Polym Sci*, **131**.

30. Gargama, H., Thakur, A. K., and Chaturvedi, S. K. (2015). Poly(vinylidene fluoride)/nickel composite materials for charge storing, electromagnetic interference absorption, and shielding applications. *J Appl Phys*, **117**.

31. Goncalves, R., Lopes, A. C., Botelho, G., Neves, I. C., and Lanceros-Mendez, S. (2013). Influence of solvent properties on the electrical response of poly(vinylidene fluoride)/NaY composites. *J Polym Res*, **20**.

32. Gonçalves, R., Martins, P. M., Caparrós, C., Martins, P., Benelmekki, M., Botelho, G., Lanceros-Mendez, S., Lasheras, A., Gutiérrez, J., and Barandiarán, J. M. (2013). Nucleation of the electroactive β-phase, dielectric and magnetic response of poly(vinylidene fluoride) composites with Fe_2O_3 nanoparticles. *J Non-Crystall Solids*, **361**, 93–99.

33. Guan, F. X., Pan, J. L., Wang, J., Wang, Q., and Zhu, L. (2010). Crystal orientation effect on electric energy storage in poly(vinylidene fluoride-co-hexafluoropropylen) copolymers. *Macromolecules*, **43**, 384–392.

34. Guan, F. X., Wang, J., Pan, J. L., Wang, Q., and Zhu, L. (2010). Effects of polymorphism and crystallite size on dipole reorientation in poly(vinylidene fluoride) and its random copolymers. *Macromolecules*, **43**, 6739–6748.

35. Hess, C. M., Rudolph, A. R., and Reid, P. J. (2015). Imaging the effects of annealing on the polymorphic phases of poly(vinylidene fluoride). *J Phys Chem B*, **119**, 4127–4132.

36. Hou, X. Y., Ng, S. C., Zhang, J., and Chang, J. S. (2015). Polymer nanocomposite dielectric based on P(VDF-TrFE)/PMMA/$BaTiO_3$ for TIPs-pentacene OFETs. *Org Electron*, **17**, 247–252.

37. Hu, P., Song, Y., Liu, H., Shen, Y., Lin, Y., and Nan, C.-W. (2013). Largely enhanced energy density in flexible P(VDF-TrFE) nanocomposites by surface-modified electrospun $BaSrTiO_3$ fibers. *J Mater Chem A*, **1**, 1688.

38. Huan, Y., Liu, Y. Y., and Yang, Y. F. (2007). Simultaneous stretching and static electric field poling of poly(vinylidene fluoride-hexafluoropropylene) copolymer films. *Polym Eng Sci*, **47**, 1630–1633.

39. Huang, W. W., Edenzon, K., Fernandez, L., Razmpour, S., Woodburn, J., and Cebe, P. (2010). Nanocomposites of poly(vinylidene fluoride) with multiwalled carbon nanotubes. *J Appl Polym Sci*, **115**, 3238–3248.

40. Hwang, S. K., Bae, I., Cho, S. M., Kim, R. H., Jung, H. J., and Park, C. (2013). High performance multi-level non-volatile polymer memory with solution-blended ferroelectric polymer/high-k insulators for low voltage operation. *Adv Funct Mater*, **23**, 5484–5493.

41. Jonscher, A. K. and Menegotto, J. (2000). Low-loss behavior of alpha-PVDF. *IEEE Trans Dielectr Electr Insul*, **7**, 303–305.

42. Khan, M. A., Bhansali, U. S., Almadhoun, M. N., Odeh, I. N., Cha, D., and Alshareef, H. N. (2014). High-performance ferroelectric memory based

on phase-separated films of polymer blends. *Adv Funct Mater*, **24**, 1372–1381.

43. Khodaparast, P. and Ounaies, Z. (2013). On the impact of functionalization and thermal treatment on dielectric behavior of low content TiO_2/PVDF nanocomposites. *IEEE Trans Dielectr Electr Insul*, **20**, 166–176.

44. Kuang, X. W., Gao, Q., and Zhu, H. (2013). Effect of calcination temperature of TiO_2 on the crystallinity and the permittivity of PVDF-TrFE/TiO_2 composites. *J Appl Polym Sci*, **129**, 296–300.

45. Kuang, X. W., Lin, S., and Zhu, H. (2013). Synthesis and characterization of poly(vinylidene-trifluoroethylene)/Ni-TiO_2. *Phys Status Solidi A*, **210**, 570–573.

46. Li, J. J., Meng, Q. J., Li, W. J., and Zhang, Z. C. (2011). Influence of crystalline properties on the dielectric and energy storage properties of poly(vinylidene fluoride). *J Appl Polym Sci*, **122**, 1659–1668.

47. Liang, C. L., Mai, Z. H., Xie, Q., Bao, R. Y., Yang, W., Xie, B. H., and Yang, M. B. (2014). Crystallization kinetics of gamma phase poly(vinylidene fluoride) (PVDF) induecd by tetrabutylammonium bisulfate. *J Polym Res*, **21**.

48. Lukyanchuk, I., Sharma, P., Nakajima, T., Okamura, S., Scott, J. F., and Gruverman, A. (2014). High-symmetry polarization domains in low-symmetry ferroelectrics. *Nano Lett*, **14**, 6931–6935.

49. Mao, Y. P., Mao, S. Y., Ye, Z. G., Xie, Z. X., and Zheng, L. S. (2010). Size-dependences of the dielectric and ferroelectric properties of $BaTiO_3$/poly(vinylidene fluoride) nanocomposites. *J Appl Phys*, **108**.

50. Martins, P., Caparros, C., Goncalves, R., Martins, P. M., Benelmekki, M., Botelho, G., and Lanceros-Mendez, S. (2012). Role of nanoparticle surface charge on the nucleation of the electroactive beta-poly(vinylidene fluoride) nanocomposites for sensor and actuator applications. *J Phys Chem C*, **116**, 15790–15794.

51. Martins, P. M., Martins, P., Correia, V., Rocha, J. G., and Lanceros-Mendez, S. (2015). Gd_2O_3:Eu nanoparticle-based poly(vinylidene fluoride) composites for indirect X-ray detection. *J Electron Mater*, **44**, 129–135.

52. Mendes, S. F., Costa, C. M., Sencadas, V., Nunes, J. S., Costa, P., Gregorio, R., and Lanceros-Mendez, S. (2009). Effect of the ceramic grain size and concentration on the dynamical mechanical and dielectric behavior of poly(vinilidene fluoride)/Pb(Zr0.53Ti0.47)O-3 composites. *Appl Phys A: Mater Sci Process*, **96**, 899–908.

53. Mendes, S. F., Costa, C. M., Caparros, C., Sencadas, V., and Lanceros-Mendez, S. (2012). Effect of filler size and concentration on the structure and properties of poly(vinylidene fluoride)/$BaTiO_3$ nanocomposites. *J Mater Sci*, **47**, 1378–1388.

54. Niu, Y. J., Yu, K., Bai, Y. Y., and Wang, H. (2015). Enhanced dielectric performance of $BaTiO_3$/PVDF composites prepared by modified process for energy storage applications. *IEEE Trans Ultrason Ferroelectr Freq Control*, **62**, 108–115.

55. Oliveira, F., Leterrier, Y., Manson, J. A., Sereda, O., Neels, A., Dommann, A., and Damjanovic, D. (2014). Process influences on the structure, piezoelectric, and gas-barrier properties of PVDF-TrFE copolymer. *J Polym Sci Part B: Polym Phys*, **52**, 496–506.

56. Olmos, D., Montero, F., Gonzalez-Gaitano, G., and Gonzalez-Benito, J. (2013). Structure and morphology of composites based on poly(vinylidene fluoride) filled with $BaTiO_3$ submicrometer particles: effect of processing and filler content. *Polym Compos*, **34**, 2094–2104.

57. Ozkazanc, E. and Guney, H. Y. (2009). The variation of the dielectric constant and loss index with temperature and draw ratio in alpha-PVDF. *J Appl Polym Sci*, **112**, 2482–2485.

58. Pal, K., Kang, D. J., Zhang, Z. X., and Kim, J. K. (2010). Synergistic effects of zirconia-coated carbon nanotube on crystalline structure of poly(vinylidene fluoride) nanocomposites: electrical properties and flame-retardant behavior. *Langmuir*, **26**, 3609–3614.

59. Patel, P. K., Yadav, K. L., and Dutta, S. (2015). Development of Ba0.95Sr0.05(Fe0.5Nb0.5)O-3/poly(vinylidene fluoride) nanocomposites for energy storage. *J Mater Sci: Mater Electron*, **26**, 4165–4171.

60. Ploss, B. and Ploss, B. (2000). Dielectric nonlinearity of PVDF-TrFE copolymer. *Polymer*, **41**, 6087–6093.

61. Ploss, B., Ploss, B., Shin, F. G., Chan, H. L. W., and Choy, C. L. (2000). Pyroelectric activity of ferroelectric PT/PVDF-TRFE. *IEEE Trans Dielectr Electr Insul*, **7**, 517–522.

62. Prabakaran, K., Mohanty, S., and Nayak, S. K. (2014). Influence of surface modified TiO_2 nanoparticles on dielectric properties of PVdF-HFP nanocomposites. *J Mater Sci: Mater Electron*, **25**, 4590–4602.

63. Rath, S. K., Dubey, S., Kumar, G. S., Kumar, S., Patra, A. K., Bahadur, J., Singh, A. K., Harikrishnan, G., and Patro, T. U. (2014). Multi-walled CNT-induced phase behaviour of poly(vinylidene fluoride) and its electro-mechanical properties. *J Mater Sci*, **49**, 103–113.

64. Sanchez, F. A., Redondo, M., and Gonzalez-Benito, J. (2015). Influence of $BaTiO_3$ submicrometric particles on the structure, morphology, and crystallization behavior of poly(vinylidene fluoride). *J Appl Polym Sci*, **132**.

65. Sun, L. L., Li, B., Zhao, Y., Mitchell, G., and Zhong, W. H. (2010). Structure-induced high dielectric constant and low loss of CNF/PVDF composites with heterogeneous CNF distribution. *Nanotechnology*, **21**.

66. Sun, L. L., Li, B., Zhao, Y., and Zhong, W. H. (2010). Suppression of AC conductivity by crystalline transformation in poly(vinylidene fluoride)/carbon nanofiber composites. *Polymer*, **51**, 3230–3237.

67. Tang, Z. C. W., Sun, L. L., Li, B., and Zhong, W. H. (2012). Structurally induced dielectric constant promotion and loss suppression for poly(vinylidene fluoride) nanocomposites. *Macromol Mater Eng*, **297**, 420–426.

68. Tong, W. S., Zhang, Y. H., Yu, L., Luan, X. L., An, Q., Zhang, Q., Lv, F. Z., Chu, P. K., Shen, B., and Zhang, Z. L. (2014). Novel method for the fabrication of flexible film with oriented arrays of graphene in poly(vinylidene fluoride-co-hexafluoropropylene) with low dielectric loss. *J Phys Chem C*, **118**, 10567–10573.

69. Wang, J., Guan, F. X., Cui, L., Pan, J. L., Wang, Q., and Zhu, L. (2014). Achieving high electric energy storage in a polymer nanocomposite at low filling ratios using a highly polarizable phthalocyanine interphase. *J Polym Sci Part B: Polym Phys*, **52**, 1669–1680.

70. Wang, Q. and Zhu, L. (2011). Polymer nanocomposites for electrical energy storage. *J Polym Sci Part B: Polym Phys*, **49**, 1421–1429.

71. Wu, S., Shao, M., Burlingame, Q., Chen, X. Z., Lin, M. R., Xiao, K., and Zhang, Q. M. (2013). A high-K ferroelectric relaxor terpolymer as a gate dielectric for organic thin film transistors. *Appl Phys Lett*, **102**.

72. Yang, L. Y., Li, X. Y., Allahyarov, E., Taylor, P. L., Zhang, Q. M., and Zhu, L. (2013). Novel polymer ferroelectric behavior via crystal isomorphism and the nanoconfinement effect. *Polymer*, **54**, 1709–1728.

73. Yu, K., Niu, Y., Xiang, F., Zhou, Y., Bai, Y., and Wang, H. (2013). Enhanced electric breakdown strength and high energy density of barium titanate filled polymer nanocomposites. *J Appl Phys*, **114**, 174107.

74. Yu, K., Niu, Y. J., Zhou, Y. C., Bai, Y. Y., and Wang, H. (2013). Nanocomposites of surface-modified $BaTiO_3$ nanoparticles filled ferroelectric polymer with enhanced energy density. *J Am Ceram Soc*, **96**, 2519–2524.

75. Yu, L. J. and Li, W. P. (2011). Stretching-induced percolation in poly(vinylidene fluoride) and nickel composites. *J Appl Polym Sci*, **120**, 2368–2373.

76. Zhang, Z. C. and Chung, T. C. M. (2007). The structure-property relationship of poly(vinylidene difluoride)-based polymers with energy storage and loss under applied electric fields. *Macromolecules*, **40**, 9391–9397.

Chapter 6

Copolymers in Dielectric Polymer Nanocomposites

Bin Li[a] **and Wei-Hong Zhong**[b]
[a]*Department of Mechanical Engineering, Wichita State University, 1845 Fairmount Street, Wichita, Kansas 67260-0133, USA*
[b]*School of Mechanical and Materials Engineering, Washington State University, 405 NE Spokane Street, Pullman, Washington 99164-2920, USA*
bin.li@wichita.edu, katie_zhong@wsu.edu

In Chapter 5, we discussed poly(vinylidene fluoride) (PVDF) as a popular matrix material for nanocomposites with high dielectric performances. In particular, because of its high polarization, piezoelectricity, and pyroelectricity, ferroelectric β-phase has been extensively studied. β-phase is widely considered the most useful PVDF structures contributing to the high energy storage and sensitivity to external stimuli. However, the crystal structure in PVDF also accounts for the biggest disadvantage of PVDF homopolymers: the high dielectric loss caused by dipole switching. This has been the major obstacle to the practical applications of PVDF-based materials in related energy and electronic applications.

Polymer Nanocomposites for Dielectrics
Edited by Wei-Hong Zhong and Bin Li

ISBN 978-981-4745-53-6 (Hardcover), 978-981-4745-54-3 (eBook)
www.panstanford.com

The stability of β-phase is another critical issue for the applications of PVDF homopolymers. In PVDF homopolymer, β-phase is not thermodynamically stable. It is usually obtained by mechanically stretching or electrically poling PVDF films prepared by compression molding or solution casting. Phase transformation to nonpolar and paraelectric α-phase often occurs with an increase in temperature. At the same time, for the applications involving substrates, such as tactile sensors and nonvolatile memories, mechanical stretching, in spite of its convenience, is not feasible for obtaining large portions of β-phase in PVDF homopolymers.

To overcome these issues, an effective approach is to introduce copolymer structures to PVDF [5, 13, 16, 21, 27, 41]. As early as 1980s, it was found that poly(vinylidene-co-trifluoroethylene) (P(VDF-TrFE)) had shown piezoelectric and ferroelectric properties superior to those of PVDF homopolymer [5]. Since then, not only the piezoelectric and pyroelectric properties of PVDF copolymers but also their energy storage properties have been extensively studied. Conventional PVDF copolymers include P(VDF-co-TrFE) [2–4, 7, 9, 18, 22, 23, 27–32, 34–36, 39, 40, 46], poly(vinylidene fluoride-co-hexafluoropropylene)(P(VDF-co-HFP)) [8, 10, 11, 13, 15, 26, 33, 37, 41, 43, 48] and the terpolymers of P(VDF-co-TrFE) [1, 43, 45]. In addition to TrFE and HFP, comonomers, such as chlorotrifluoroethylene (CTFE), chlorofuoroethylene (CFE), and bromotrifluoroethylene, have also been applied to synthesize PVDF copolymers [12, 16, 17].

The copolymerization of VDF with these comonomers can create smaller crystals and favor the formation of β-phase [30], leading to more stable polar structures with reduced energy loss. However, PVDF copolymers often possess more structural defects, and their cost is much higher than that of PVDF and other polymer dielectric films, which prevents the practical applications of these copolymer materials. This chapter will review the recent achievements in the research on these PVDF copolymers, in particular the copolymers with TrFE and HFP, and their nanocomposites, for dielectric energy storage applications. The discussion on the effects of nanomaterials on crystal structures of the copolymers will be included as well.

6.1 Polarization Mechanisms in PVDF and PVDF Copolymers

Dielectric polarization concerns the formation and orientation of dipole moments in dielectric materials. For dielectric materials like PVDF, they possess spontaneous dipole moments, as a result of highly polar chemical and crystal structures. The dipole orientation, or dipole switching phenomena upon charging and discharging processes, is critical to energy storage performance. In previous studies on dipole orientation behavior of PVDF and its copolymers [13, 15, 16], a model of dipole orientation under electrical field was proposed, as shown in Fig. 6.1. The applied electric field aligns the dipoles in the ferroelectric crystal, i.e., β-phase in PVDF. This alignment of dipole moments subsequently induces a compensation polarization in the amorphous phase at the interface between crystal region and amorphous region. This interfacial compensation polarization accounts for the coupling between different ferroelectric crystals. Typically, high β content and large crystal size favor this coupling.

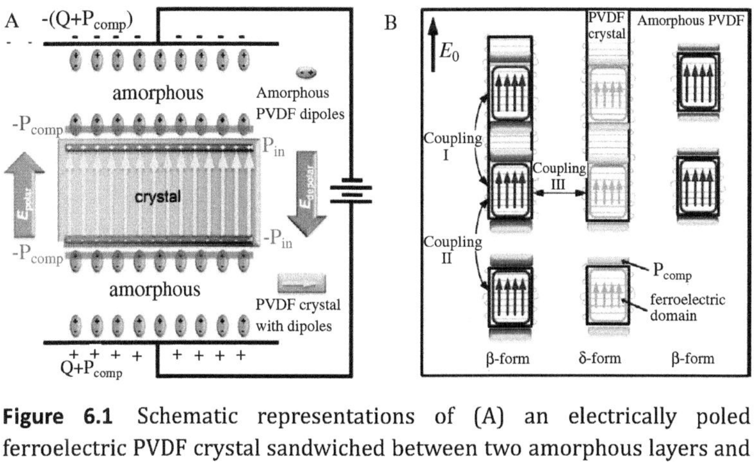

Figure 6.1 Schematic representations of (A) an electrically poled ferroelectric PVDF crystal sandwiched between two amorphous layers and (B) many electrically poled ferroelectric PVDF crystals in an amorphous matrix, showing different coupling interactions among ferroelectric domains. The chain direction in the lamellar crystals is perpendicular to the film normal (or the external electric field). Reprinted with permission from Ref. [14], Copyright 2010, American Chemical Society.

It was found that the dipoles could easily orient when there is a strong coupling effect between different ferroelectric crystals, as shown in Fig. 6.1 [14], which favors higher polarization and higher energy storage density. However, according to Fig. 2.9 in Chapter 2, we know that, for dielectric capacitor applications, it is insufficient to only have high energy storage density. High discharged energy density is also desired. This suggests that the oriented/polarized dipoles should be able to switch back to its initial random state in order to obtain high discharged energy density. While strong coupling favors the dipole orientation during charging, it disfavors the inverse processes during discharging.

A feasible way is to reduce the crystal size to nanoscale and weaken the coupling between the polar crystal structures. Figures 6.2 and 6.3 present the crystal morphologies of both homopolymer and copolymers of PVDF [19]. In PVDF homopolymer without stretching, we can observe very large PVDF spherulites with an average diameter around 5 μm. In contrast, in the copolymers, both dicopolymers and tercopolymers show needle-like crystal structures with a much smaller crystal size. The length of these needles are no larger than 200 nm. Meanwhile, a lot more voids are observed in both copolymer structures. After the copolymer structures were introduced, these changes in crystal structures, such as crystal size and morphologies, suggest the potential weaker coupling between ferroelectric crystals in these copolymers.

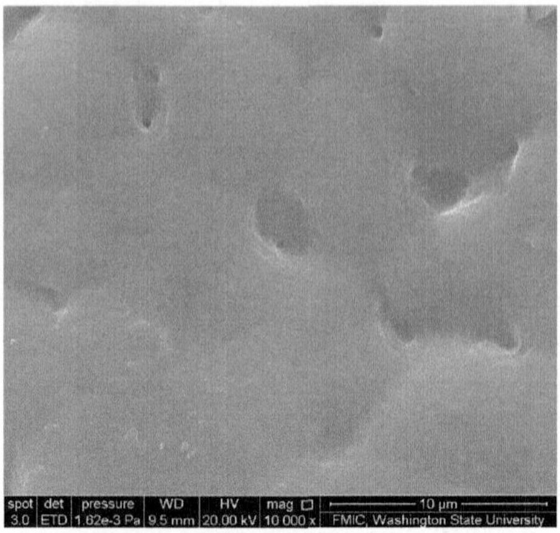

Figure 6.2 SEM image of morphology of crystal structures of PVDF homopolymer prepared by solution casting.

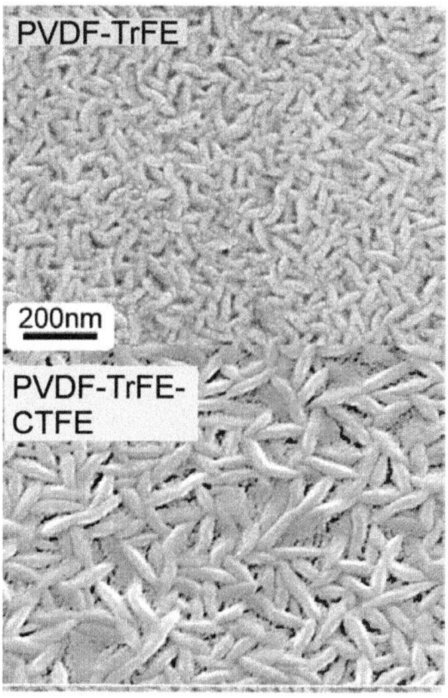

Figure 6.3 SEM image of morphologies of crystal structures of P(VDF-TrFE) (75/25) and P(VDF-TrFE-CTFE) prepared by spin coating. Reprinted from Ref. [19], with permission from John Wiley and Sons.

6.2 Dicopolymers of PVDF and Their Nanocomposites

P(VDF-TrFE) is one of the most researched copolymers of PVDF. In Fig. 6.4, both crystal structures and dielectric polarization behavior of P(VDF-TrFE) are presented [5]. In this copolymer, the molar ratio between VDF and TrFE is 72:28. Both PVDF and P(VDF-TrFE) films were spin-coated on the aluminum-coated silicon and annealed at 135°C. According to XRD results, both polymers had β-crystal. However, it is obvious that the P(VDF-TrFE) has a much higher crystallinity of β-phase in the polymer matrix. Meanwhile, the different crystal morphologies were also observed. In PVDF the tiny granular crystal structure was dominant, while in the copolymer the rod-like crystal structure prevailed with many cracks and voids. Regarding the cracking issues, the volume shrinkage inside the copolymer crystal structures was proposed as the main mechanism. The differences in crystal structures contributed to completely different dielectric properties as well. Obviously, P(VDF-TrFE) exhibited both high energy storage density and reduced energy loss. However, the same research also found reduced breakdown strength of the copolymer (402 ± 25 MV/m), compared with that of PVDF homopolymer (524 ± 25 MV/m), which has been an issue in PVDF copolymers [6].

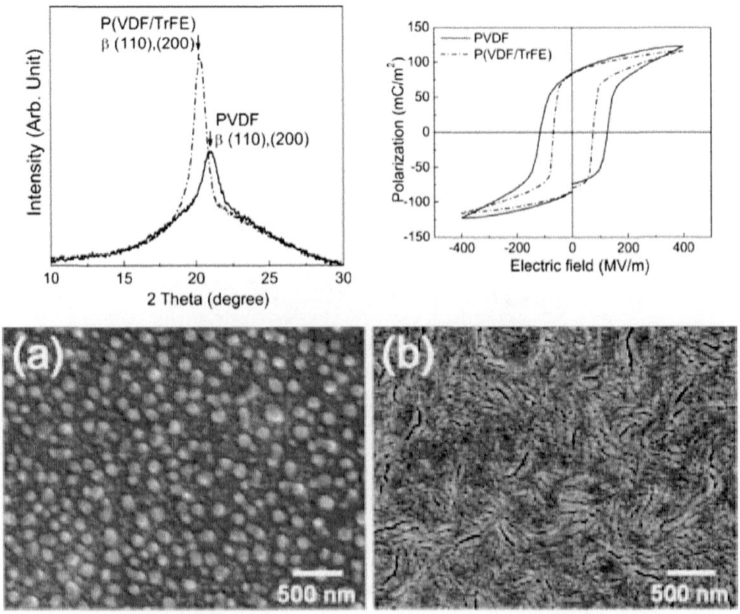

Figure 6.4 XRD patterns, hysteresis loops, and morphologies of PVDF (a) and P(VDF-TrFE) (b) on aluminum-coated silicon. Reprinted from Ref. [5], with permission from John Wiley and Sons.

The molar ratio is very critical to the desired crystal structures and dielectric performances of the PVDF copolymers. In the study shown in Fig. 6.4 [5], there is only one molar ratio between two monomers. A theoretical study of P(VDF-TrFE) copolymers was shown in Fig. 6.5 [3]. This study was intended to analyze the molecular mechanism of the polarization switching in PVDF and (P(VDF–TrFE)) using semi-empirical and ab initio quantum chemical methods. Two different molar ratios between VDF and TrFE were studied. It is probably because this theory only considered the molecular chains without taking into consideration the polymorphism of PVDF and its copolymers, the calculated results showed increased hysteresis loss in the copolymers compared with pure PVDF. However, it is very clear that the molar ratio between two monomers affects the hysteresis behaviors of the copolymer structures. Currently, the understanding of the effects of the ratios between different monomers is still insufficient, and most research has only focused on a fixed molar ratio.

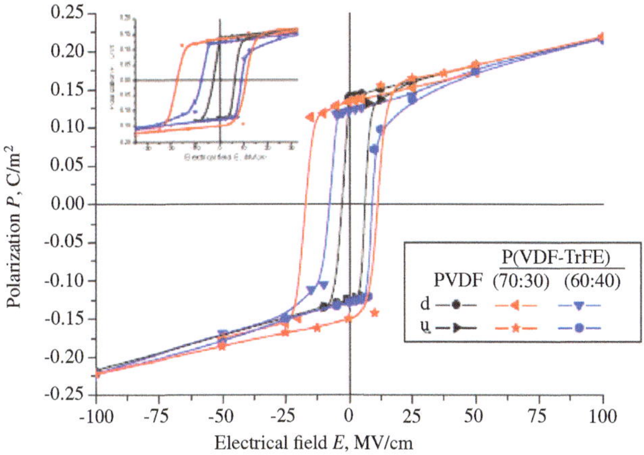

Figure 6.5 Calculated data of hysteresis loop for different TrFE contents in the molecular polymer chain: wide region of calculated hysteresis loop. The inset shows the central region of the hysteresis loop only. Coercive fields are ~5 $MVcm^{-1}$ for pure PVDF, ~8 $MVcm^{-1}$ for 40%, and more than ~10 MV cm^{-1} for 30% of TrFE in P(VDF–TrFE). Reproduced with permission from Ref. [3], Copyright 2007, IOP Publishing. All rights reserved.

Similar to the research on PVDF nanocomposites introduced in Chapter 5, although the significance of polymorphism of PVDF copolymers to dielectric properties has been well accepted, the understanding on the effects of nanomaterials on the polymorphism of the copolymers is very limited. In most reported work, the focus was still only on the dispersion and distribution of nanomaterials as well as interfacial interactions. The synergistic effects between these factors and crystal structures of the copolymers on dielectric properties were rarely explored.

Figure 6.6 presents the XRD patterns and dielectric properties of P(VDF-TrFE)/Ni-TiO_2 nanocomposites[23]. According to the XRD patterns, in P(VDF-TrFE), the peaks at 20.00°, 35.30°, and 41.02° are the characteristic signals of the β ferroelectric phase of P(VDF-TrFE). The addition of Ni-TiO_2 nanoparticles seemed to decrease the crystallinity of the copolymer, as evidenced by the less intense and broader diffraction peaks of the copolymer. The dielectric properties of the resulting nanocomposites exhibited a clear concentration dependence. With the increasing loading, the dielectric permittivity increased accordingly. Meanwhile, it is interesting to notice that the nanocomposites showed a lower dielectric loss compared with the pure polymer. The study revealed the reduced crystallinity of β-phase in the copolymer and the increase in dielectric permittivity when the content of Ni-TiO_2 nanoparticle increased. It implied that in this nanocomposite, the dispersion of nanomaterials and interfacial polarization might be the dominating factors in dielectric properties.

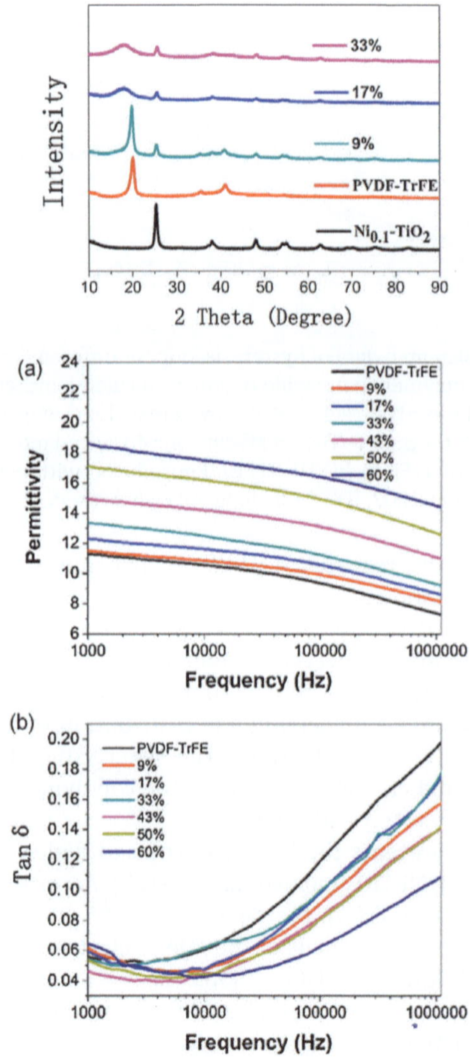

Figure 6.6 XRD patterns and dielectric properties of P(VDF-TrFE)/Ni-TiO_2 nanocomposites. Reprinted from Ref. [23], with permission from John Wiley and Sons.

The same research group reported a similar nanocomposite system, as shown in Fig. 6.7. In this study [22], two different TiO_2 nanoparticles were prepared at different calcination temperatures. They were incorporated to P(VDF-TrFE) to fabricate nanocomposites. It was found both TiO_2 nanoparticles could lead to the increase of crystallinity of β-phase in the copolymer. In particular, the highest crystallinity was found in CT850 nanocomposites, in which the TiO_2 nanoparticles were calcined at 850°C. The dielectric properties

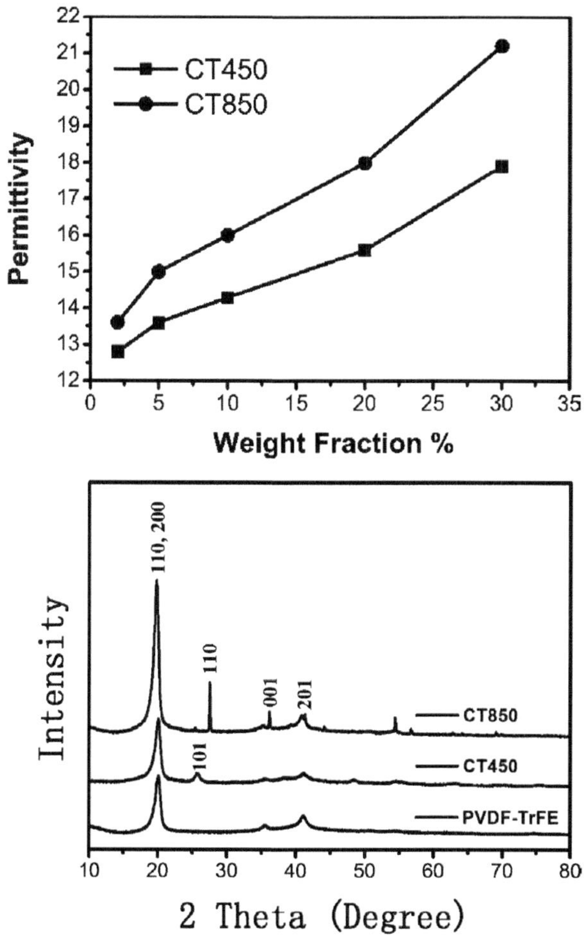

Figure 6.7 Dielectric properties and XRD spectra of P(VDF-TrFE)/TiO_2 nanocomposites. CT450, calcination temperature 450°C; CT850, calcination temperature 850°C. Reprinted from Ref. [22], with permission from John Wiley and Sons.

of the resulting nanocomposites increased with the loading of nanoparticles. CT850 nanocomposites showed higher dielectric constant than CT450 nanocomposites. Both studies conducted by the same research group used the same P(VDF-TrFE), in which the molar ratio between VDF and TrFE was 70/30. However, the effects of nanomaterials on the crystal structures were very different, although they are all TiO_2-based nanomaterials. Both composition and nanostructures determined by processing conditions had a great impact on the crystal structures of the copolymer. By comparing these two studies, it is clear that CT850 nanocomposites had the highest dielectric permittivity among the three groups of nanocomposites in Figs. 6.6 [23] and 6.7 [22]. This comparison reveals the contribution of the polymorphism and dielectric properties of the copolymer matrix and possible synergy between the nanomaterials and the polymer matrix.

P(VDF-HFP) is another popular copolymer of PVDF. In a study shown in Fig. 6.8 [26], it was modified by hydantoin epoxy resin coated $BaTiO_3$ nanoparticles. Pure P(VDF-HFP) showed an amorphous structure in this study, and the addition of $BaTiO_3$ nanoparticles did not change the phase structures of the copolymers. Therefore, this work focused only on the dispersion of nanomaterials and their interface. Obviously, the addition of the modified BaTiO3 nanoparticles could effectively increase the energy density of the nanocomposites. Meanwhile, reduced energy efficiency was obtained, suggesting high energy loss.

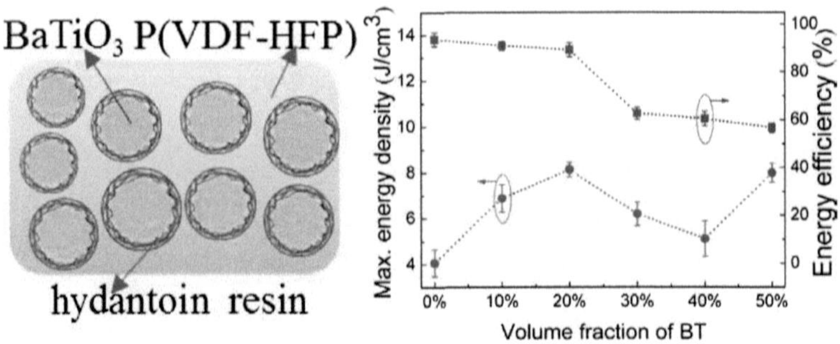

Figure 6.8 P(VDF-HFP)/hydantoin epoxy resin modified $BaTiO_3$ nanocomposites and their energy storage performances. Reprinted with permission from Ref. [26], Copyright 2015, American Chemical Society.

In the study of P(VDF-HFP)/reduced graphene oxide (rGO) nanocomposites (Fig. 6.9) [38], the pure P(VDF-HFP) possessed an α-phase structure. The addition of uniformly dispersed rGO nanoparticles did not affect the phase structure of P(VDF-HFP), and the crystallinity of the matrix was slightly reduced by the fully exfoliated rGO nanoparticles. The resulting nanocomposites showed a clear dependence on the concentration of rGO. The maximum dielectric constant was achieved with 0.84 vol% rGO. The decrease in dielectric properties at 1.00 vol% may be caused by the increasing number of agglomerates in the nanocomposites, which weakened the interfacial polarization.

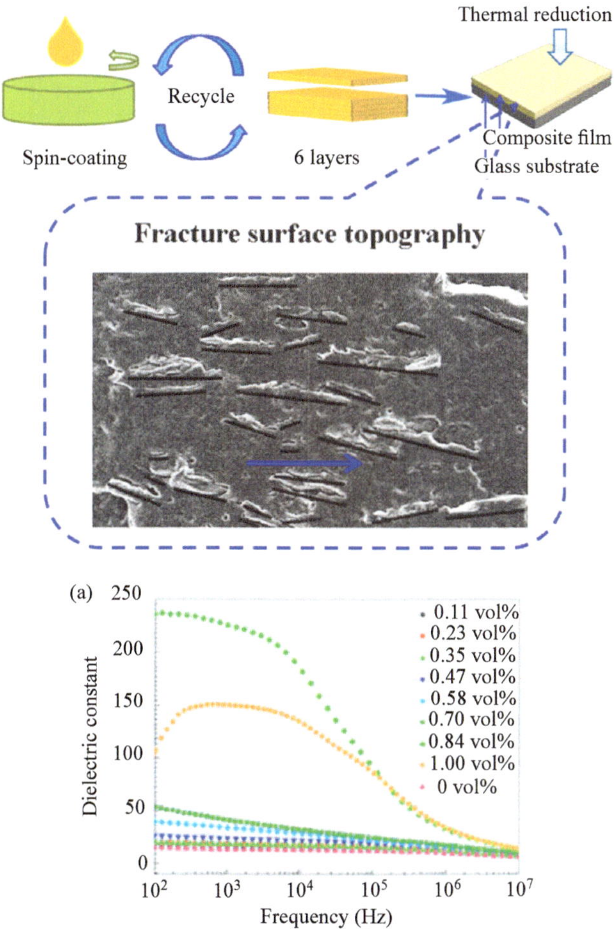

Figure 6.9 Fabrication of P(VDF-HFP)/oriented reduced graphene oxide (rGO) nanocomposites and their dielectric property. Reprinted with permission from Ref. [38], Copyright 2014, American Chemical Society.

Kaolinite clay nanomaterials were added to P(VDF-HFP) to improve energy performance (Fig. 6.10) [37]. In this study, the addition of clay nanomaterials did not show noticeable contributions to the β-phase crystal of the copolymer, in neither the unstretched nor the stretched nanocomposite films. The existence of uniformly dispersed clay nanomaterials in the matrix led to a remarkable increase in breakdown strength, showing nanocomposites of great potential in high electric field applications. Because of this dramatic improvement in breakdown strength, the high discharged energy density was obtained. Meanwhile, the high energy efficiency was achieved in this group of nanocomposites. As discussed in Section 6.1, the reason for the high loss was that oriented dipoles could not switch back to their initial random state in the polar crystal structures. In this study, the researcher found that the addition of

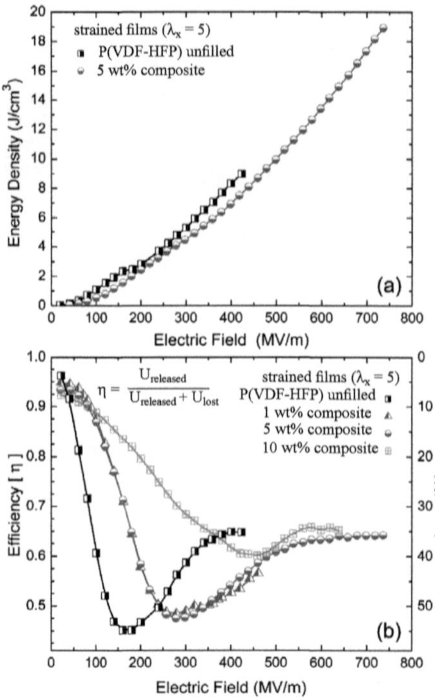

Figure 6.10 (a) Discharged energy density calculated from the *D–E* hysteresis loops: comparing the unfilled stretched P(VDF-HFP) films against its 5 wt% kaolinite nanocomposite. (b) Energy efficiencies for stretched films of unfilled P(VDF-HFP) and its 1, 5, and 10 wt% nanocomposites. Reprinted with permission from Ref. [37], Copyright 2011, AIP Publishing LLC.

kaolinite clay nanomaterials did not promote the formation of polar β-crystal structures, as shown in Fig. 6.11 [37]. In other words, the nonpolar α-phase was stabilized by the kaolinite clay nanomaterials, which was considered as the mechanism for the enhanced energy efficiency.

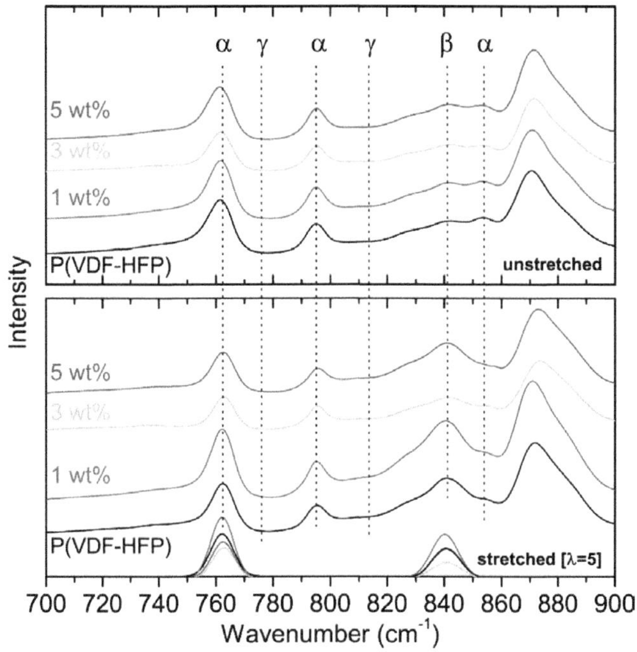

Figure 6.11 FT-IR spectra of P(VDF-HFP) copolymer and its 1, 3, and 5 wt% nanocomposites: attenuated total reflection intensities from stretched (bottom) and as-cast (top, unstretched) films. The c-exclusive absorptions at 776 and 812 cm^{-1} are absent in all systems. For the stretched films, the background-corrected fits for the a-764 cm^{-1} and b-840 cm^{-1} absorption peaks are also shown. Reprinted with permission from Ref. [37], Copyright 2011, AIP Publishing LLC.

6.3 Tercopolymers of PVDF

In the beginning of this chapter, a couple of advantages of PVDF copolymers were introduced. However, the superiority of dicopolymers of PVDF to PVDF homopolymers are often not very obvious. In order to bring the performance of PVDF-based materials to a whole new level and achieve practical applications, the terpolymers of PVDF have been gaining more and more attention [1, 16, 24, 42]. By introducing third bulky monomers to the dicopolymers, nano-size crystal structures can be realized. More often, the terpolymers of PVDF exhibit a relaxor-type ferroelectric behavior, as shown in Fig. 2.9 in Chapter 2, contributing to both high energy storage density and high discharged energy density.

Figure 6.12 summarized the hysteresis behavior of PVDF homopolymer and various dicopolymers and terpolymers. Obviously, both chemical structures and compositions could remarkably affect the energy storage performances. The terpolymer P(VDF-TrFE-CTFE) exhibited high polarization, while its energy loss, indicated by the loop area, is very low and the hysteresis showed relaxor-type ferroelectricity.

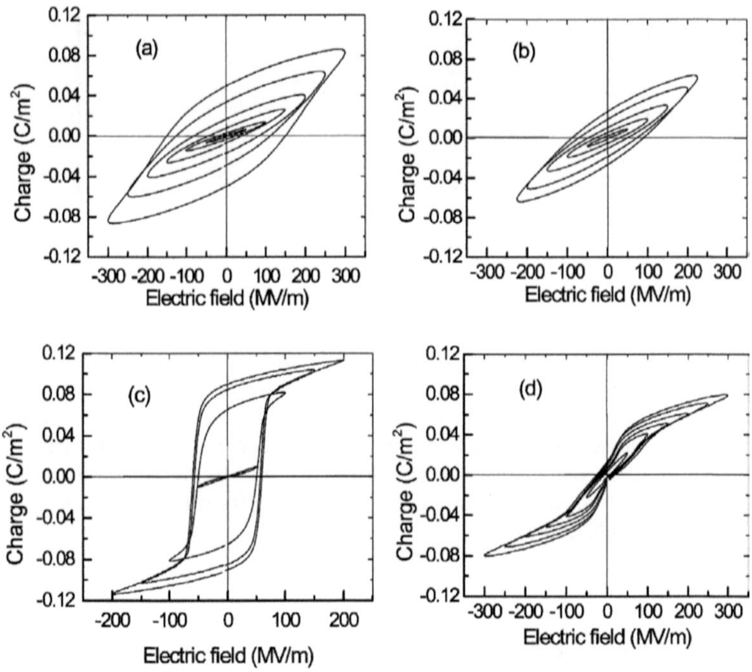

Figure 6.12 Bipolar *D–E* hysteresis loops (before breakdown) of (a)PVDF homopolymer, (b) VDF/CTFE (92.2/7.8 mol%) copolymer, (c) VDF/TrFE (63.3/36.7 mol%) copolymer , and (d) VDF/TrFE/CTFE (65.6/26.7/7.7 mol%) terpolymer. Reprinted with permission from Ref. [47], Copyright 2007, American Chemical Society.

A more in-depth study was conducted to investigate the effects of chemical compositions on the ferroelectric behaviors of both dicopolymers and terpolymers, as shown in Fig. 6.13 [46]. P(VDF-TrFE-CTFE) is superior to both P(VDF-CTFE) and P(VDF-TrFE) copolymers. In addition, the P(VDF-TrFE-CTFE) terpolymer has a Curie transition at the near-ambient temperature. Therefore, the terpolymer is more suitable for high-power capacitors with high energy density and low energy loss. Compared with CTFE, the increase in TrFE content had fewer effects on the hysteresis behaviors of the copolymers.

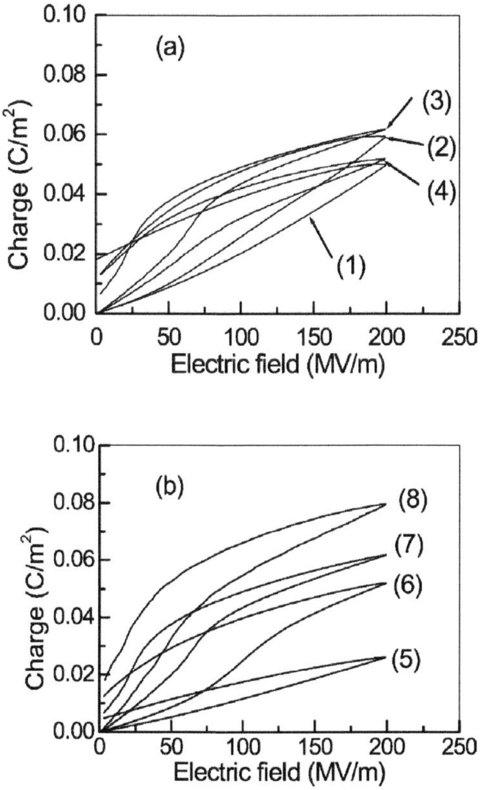

Figure 6.13 Comparison of unipolar *D–E* hysteresis curves of two sets of VDF/TrFE/CTFE terpolymers that contain (a) a fixed CTFE content (7.5 mol%) and various VDF/TrFE mole ratios, including (1) 92.2/0, (2) 80.7/11.6, (3) 65.6/26.7, and (4) 58.4/34.2 mol%, and (b) a fixed VDF content (65 mol%) and various TrFE/CTFE mole ratios, including (5) 36.7/0,(6) 27.2/5.5, (7) 26.7/7.7, and (8) 22.5/9.3 mol%. Reprinted with permission from Ref. [47], Copyright 2007, American Chemical Society.

Because of the high cost and availability of the terpolymers of PVDF, the research on their nanocomposites is far less than sufficient [20, 25, 44], although the superiority of these terpolymers is remarkable. Only very few nanomaterials have been applied to these terpolymers of PVDF, showing property improvement to different levels. However, the nanomodification did not seem to fully realize the potential of this group of relaxor-type ferroelectric polymer materials. This is probably due to the lack of understanding on dicopolymer-based nanocomposites. Knowledge on the synergy between the polymorphs of copolymers and nanomaterials, and the interactions between different nanomaterials and different monomer structures is rarely understood. The molar ratio of different monomers and their arrangement on the polymer chain structures are essential to the dielectric properties as well. Systematic and in-depth studies on these issues are still needed.

References

1. Bauer, F. (2012). Review on the properties of the ferrorelaxor polymers and some new recent developments. *Appl Phys A: Mater Sci Process*, **107**, 567–573.
2. Bharti, V., Xu, H., Cheng, Z. Y., Mai, T., and Zhang, Q. M. (2001). Quantitative analysis of structural, relaxational and electrostrictive properties of PVDF-TRFE/PMMA films irradiated with high-energy electrons. *IEEE Trans Dielectr Electr Insul*, **8**, 718–724.
3. Bystrov, V. S., Bystrova, N. K., Paramonova, E. V., Vizdrik, G., Sapronova, A. V., Kuehn, M., Kliem, H., and Kholkin, A. L. (2007). First principle calculations of molecular polarization switching in P(VDF-TrFE) ferroelectric thin Langmuir-Blodgett films. *J Physics: Condense Matter*, **19**.
4. Chan, H. L. W., Zhang, Q. Q., Ng, W. Y., and Choy, C. L. (2000). Dielectric permittivity of PCLT/PVDF-TRFE nanocomposites. *IEEE Trans Dielectr Electr Insul*, **7**, 204–207.
5. Chen, S. T., Yao, K., Tay, F. E. H., and Chew, L. L. S. (2010). Comparative investigation of the structure and properties of ferroelectric poly(vinylidene fluoride) and poly(vinylidene fluoride-trifluoroethylene) thin films crystallized on substrates. *J Appl Polym Sci*, **116**, 3331–3337.
6. Chen, X. Z., Cheng, Z. X., Liu, L., Yang, X. D., Shen, Q. D., Hu, W. B., and Li, H. T. (2013). Evolution of nanopolar phases, interfaces, and increased

dielectric energy storage capacity in photoinitiated cross-linked poly(vinylidene fluoride)-based copolymers. *Colloid Polym Sci*, **291**, 1989–1997.

7. Chen, X. Z., Li, X. Y., Qian, X. S., Wu, S., Lu, S. G., Gu, H. M., Lin, M. R., Shen, Q. D., and Zhang, Q. M. (2013). A polymer blend approach to tailor the ferroelectric responses in P(VDF-TrFE) based copolymers. *Polymer*, **54**, 2373–2381.

8. Ehrhardt, C., Fettkenhauer, C., Glenneberg, J., Munchgesang, W., Leipner, H. S., Wagner, G., Diestelhorst, M., Pientschke, C., Beige, H., and Ebbinghaus, S. G. (2014). Enhanced dielectric properties of sol-gel-$BaTiO_3$/P(VDF-HFP) composite films without surface functionalization. *RSC Adv*, **4**, 40321–40329.

9. El Shafee, E., El Gamal, M., and Isa, M. (2012). Electrical properties of multi walled carbon nanotubes/poly(vinylidene fluoride/trifluoroethylene) nanocomposites. *J Polym Res*, **19**.

10. Feng, Y., Li, W. L., Hou, Y. F., Yu, Y., Cao, W. P., Zhang, T. D., and Fei, W. D. (2015). Enhanced dielectric properties of PVDF-HFP/$BaTiO_3$-nanowire composites induced by interfacial polarization and wire-shape. *J Mater Chem C*, **3**, 1250–1260.

11. Frubing, P., Wang, F. P., and Wagener, M. (2012). Relaxation processes and structural transitions in stretched films of poly(vinylidene fluoride) and its copolymer with hexafluoropropylene. *Appl Phys A: Mater Sci Process*, **107**, 603–611.

12. Gadinski, M. R., Han, K., Li, Q., Zhang, G. Z., Reainthippayasakul, W., and Wang, Q. (2014). High energy density and breakdown strength from beta and gamma phases in poly(vinylidene fluoride-co-bromotrifluoroethylene) copolymers. *ACS Appl Mater Interfaces*, **6**, 18981–18988.

13. Guan, F. X., Pan, J. L., Wang, J., Wang, Q., and Zhu, L. (2010). Crystal orientation effect on electric energy storage in poly(vinylidene fluoride-co-hexafluoropropylen) copolymers. *Macromolecules*, **43**, 384–392.

14. Guan, F. X., Wang, J., Pan, J. L., Wang, Q., and Zhu, L. (2010). Effects of polymorphism and crystallite size on dipole reorientation in poly(vinylidene fluoride) and its random copolymers. *Macromolecules*, **43**, 6739–6748.

15. Guan, F. X., Wang, J., Pan, J. L., Wang, Q., and Zhu, L. (2011). Time and poling history dependent energy storage and discharge behaviors in poly(vinylidene fluoride-co-hexafluoropropylene) random copolymers. *Chin J Polym Sci*, **29**, 65–80.

16. Guan, F. X., Yang, L. Y., Wang, J., Guan, B., Han, K., Wang, Q., and Zhu, L. (2011). Confined ferroelectric properties in poly(vinylidene fluoride-co-chlorotrifluoroethylene)-graft-polystyrene graft copolymers for electric energy storage applications. *Adv Funct Mater*, **21**, 3176–3188.

17. Hilczer, B., Smogor, H., and Goslar, J. (2006). Dielectric response of polymer relaxors. *J Mater Sci*, **41**, 117–127.

18. Hu, P., Song, Y., Liu, H., Shen, Y., Lin, Y., and Nan, C.-W. (2013). Largely enhanced energy density in flexible P(VDF-TrFE) nanocomposites by surface-modified electrospun $BaSrTiO_3$ fibers. *J Mater Chem A*, **1**, 1688.

19. Hwang, S. K., Bae, I., Cho, S. M., Kim, R. H., Jung, H. J., and Park, C. (2013). High performance multi-level non-volatile polymer memory with solution-blended ferroelectric polymer/high-k insulators for low voltage operation. *Adv Funct Mater*, **23**, 5484–5493.

20. Javadi, A., Xiao, Y., Xu, W., and Gong, S. (2012). Chemically modified graphene/P(VDF-TrFE-CFE) electroactive polymer nanocomposites with superior electromechanical performance. *J Mater Chem*, **22**, 830–834.

21. Jung, H. M., Kang, J.-H., Yang, S. Y., Won, J. C., and Kim, Y. S. (2010). Barium titanate nanoparticles with diblock copolymer shielding layers for high-energy density nanocomposites. *Chem Mater*, **22**, 450–456.

22. Kuang, X. W., Gao, Q., and Zhu, H. (2013). Effect of calcination temperature of TiO_2 on the crystallinity and the permittivity of PVDF-TrFE/TiO_2 composites. *J Appl Polym Sci*, **129**, 296–300.

23. Kuang, X. W., Lin, S., and Zhu, H. (2013). Synthesis and characterization of poly(vinylidene-trifluoroethylene)/Ni-TiO_2. *Physica Status Solidi A*, **210**, 570–573.

24. Li, H. F., Tan, K. Y., Hao, Z. M., and He, G. W. (2011). Thermal characterization of a series of poly(vinylidene fluoride-chlorotrifluoroethylene-trifluoroethylene) terpolymer films. *J Therm Anal Calorim*, **105**, 357–364.

25. Li, Q., Han, K., Gadinski, M. R., Zhang, G., and Wang, Q. (2014). High energy and power density capacitors from solution-processed ternary ferroelectric polymer nanocomposites. *Adv Mater*, **26**, 6244–6249.

26. Luo, H., Zhang, D., Jiang, C., Yuan, X., Chen, C., and Zhou, K. C. (2015). Improved dielectric properties and energy storage density of poly(vinylidene fluoride-co-hexafluoropropylene) nanocomposite with hydantoin epoxy resin coated $BaTiO_3$. *ACS Appl Mater Interfaces*, **7**, 8061–8069.

27. Lutkenhaus, J. L., McEnnis, K., Serghei, A., and Russell, T. P. (2010). Confinement effects on crystallization and curie transitions of poly(vinylidene fluoride-co-trifluoroethylene). *Macromolecules*, **43**, 3844–3850.

28. Oliveira, F., Leterrier, Y., Manson, J. A., Sereda, O., Neels, A., Dommann, A., and Damjanovic, D. (2014). Process influences on the structure, piezoelectric, and gas-barrier properties of PVDF-TrFE copolymer. *J Polym Sci Part B: Polym Phys*, **52**, 496–506.

29. Park, Y. J., Kang, S. J., Lotz, B., Brinkmann, M., Thierry, A., Kim, K. J., and Park, C. (2008). Ordered ferroelectric PVDF-TrFE thin films by high throughput epitaxy for nonvolatile polymer memory. *Macromolecules*, **41**, 8648–8654.

30. Ploss, B. and Ploss, B. (2000). Dielectric nonlinearity of PVDF-TrFE copolymer. *Polymer*, **41**, 6087–6093.

31. Ploss, B. Ploss, B., Shin, F. G., Chan, H. L. W., and Choy, C. L. (2000). Pyroelectric activity of ferroelectric PT/PVDF-TRFE. *IEEE Trans Dielectr Electr Insul*, **7**, 517–522.

32. Ploss, B., Ng, W. Y., Chan, H. L. W., Ploss, B., and Choy, C. L. (2001). Poling study of PZT/P(VDF-TrFE) composites. *Compos Sci Technol*, **61**, 957–962.

33. Prabakaran, K., Mohanty, S., and Nayak, S. K. (2014). Influence of surface modified TiO_2 nanoparticles on dielectric properties of PVDF-HFP nanocomposites. *J Mater Sci: Mater Electron*, **25**, 4590–4602.

34. Sigamani, N., Ounaies, Z., and Sodano, H. (2013). Effect of graphene oxide on the ferroelectric properties of P(VDF-TrFE) (56/44), *Proceedings of the American Society for Composites*.

35. Sorokin, A. V., Fridkin, V. M., and Ducharme, S. (2005). Pyroelectric study of polarization switching in Langmuir-Blodgett films of poly(vinylidene fluoride trifluoroethylene). *J Appl Phys*, **98**.

36. Stephanovich, V. A., Glinchuk, M. D., Kirichenko, E. V., and Hilczer, B. (2003). Theory of radiation induced relaxor behavior of poly(vinylidene fluoride-trifluoroethylene) copolymers. *J Appl Phys*, **94**, 5937–5944.

37. Tomer, V., Manias, E., and Randall, C. A. (2011). High field properties and energy storage in nanocomposite dielectrics of poly(vinylidene fluoride-hexafluoropropylene). *J Appl Phys*, **110**.

38. Tong, W. S., Zhang, Y. H., Yu, L., Luan, X. L., An, Q., Zhang, Q., Lv, F. Z., Chu, P. K., Shen, B., and Zhang, Z. L. (2014). Novel method for the fabrication of flexible film with oriented arrays of graphene in poly(vinylidene

fluoride-co-hexafluoropropylene) with low dielectric loss. *J Phys Chem C*, **118**, 10567–10573.

39. Valiyaneerilakkal, U., and Varghese, S. Poly (vinylidene fluoride-trifluoroethylene)/barium titanate nanocomposite for ferroelectric nonvolatile memory devices, 042131.

40. Valiyaneerilakkal, U., Singh, A., Singh, K., Subash, C. K., Abbas, S. M., Komaragiri, R., and Varghese, S. (2014). Ferroelectric characteristics of MFIS structure with P(VDF-TrFE)/$BaTiO_3$ nanocomposite as ferroelectric layer. *Appl Phys A: Mater Sci Process*, **117**, 1535–1540.

41. Wang, J., Guan, F. X., Cui, L., Pan, J. L., Wang, Q., and Zhu, L. (2014). Achieving high electric energy storage in a polymer nanocomposite at low filling ratios using a highly polarizable phthalocyanine interphase. *J Polym Sci Part B: Polym Phys*, **52**, 1669–1680.

42. Wang, J. W., Shen, Q. D., Bao, H. M., Yang, C. Z., and Zhang, Q. M. (2005). Microstructure and dielectric properties of P(VDF-TrFE-CFE) with partially grafted copper phthalocyanine oligomer. *Macromolecules*, **38**, 2247–2252.

43. Wegener, M., Kunstler, W., Richter, K., and Gerhard-Multhaupt, R. (2002). Ferroelectric polarization in stretched piezo- and pyroelectric poly(vinylidene fluoride-hexafluoropropylene) copolymer films. *J Appl Phys*, **92**, 7442–7447.

44. Wen, F., Xu, Z., Xia, W., Wei, X., and Zhang, Z. (2013). High dielectric permittivity and low dielectric loss nanocomposites based on poly(VDF–TrFE–CTFE) and graphene nanosheets. *J Adv Dielectr*, **03**, 1350010.

45. Wu, S., Shao, M., Burlingame, Q., Chen, X. Z., Lin, M. R., Xiao, K., and Zhang, Q. M. (2013). A high-K ferroelectric relaxor terpolymer as a gate dielectric for organic thin film transistors. *Appl Phys Lett*, **102**.

46. Zhang, L., Shan, X. B., Wu, P. X., and Cheng, Z. Y. (2012). Dielectric characteristics of $CaCu_3Ti_4O_{12}$/P(VDF-TrFE) nanocomposites. *Appl Phys A: Mater Sci Process*, **107**, 597–602.

47. Zhang, Z. C. and Chung, T. C. M. (2007). The structure-property relationship of poly(vinylidene difluoride)-based polymers with energy storage and loss under applied electric fields. *Macromolecules*, **40**, 9391–9397.

48. Zhou, Z., Carr, J., Mackey, M., Yin, K. Z., Schuele, D., Zhu, L., and Baer, E. (2013). Interphase/interface modification on the dielectric properties of polycarbonate/poly(vinylidene fluoride-co-hexafluoropropylene) multilayer films for high-energy density capacitors. *J Polym Sci Part B: Polym Phys*, **51**, 978–991.

Chapter 7

Dielectric Properties of Bionanocomposites

Guan Gong[a] **and Bin Li**[b]
[a]*Swerea SICOMP AB, P.O. Box 271, SE-941 26 Piteå, Sweden*
[b]*Department of Mechanical Engineering, Wichita State University, 1845 Fairmount Street, Wichita, Kansas 67260-0133, USA*
guan.gong@swerea.se, bin.li@wichita.edu

Bionanocomposites represent an emerging group of nanostructured hybrid materials. Besides the structural and functional properties similar to conventional nanocomposites, the inherent properties, namely, biocompatibility and biodegradability, make these environmentally benign materials highly attractive for both academia and industry. Research on bionanocomposites is an interdisciplinary field that is closely related to packaging and biological systems, such as tissue engineering and drug delivery, and also extended to electronics, sensor, and energy applications, just to name a few. This chapter will give an overview on the research progress of self-assembled bionanocomposites and artificial nanocomposites composed of typical biopolymers and nanoparticles, which exhibit

Polymer Nanocomposites for Dielectrics
Edited by Wei-Hong Zhong and Bin Li

ISBN 978-981-4745-53-6 (Hardcover), 978-981-4745-54-3 (eBook)
www.panstanford.com

dielectric functions and can have promising potential to be used in biological, electronic, and energy storage applications.

7.1 Introduction

Dwindling fossil resources and growing environmental concerns have led to increasing need for sustainable solutions. Overdependence on petroleum resources and environmental impact/carbon footprint can be alleviated by biomaterials development using biological resources or more precisely annually renewable resources. Bionanocomposites are formed by the polymer matrix and reinforcing phase, either of them or both based on renewable resources, and show at least one dimension on the nanometer scale. Bionanocomposites have been the subject of extensive research since the last two decades. These efforts have generated environment-friendly applications for many uses, such as for automotive, packaging, and household products, to name some [1]. Multifunctionality is a trend of development to expand the applications of bionanocomposites.

Dielectric bionanocomposites can be utilized for various applications such as conductive elements, heating devices, electromagnetic interference, supercapacitor, shielding and antimicrobial elements, and bio-applications like those of tissue scaffolds for the replacement or restoration of damaged or malfunctioning tissue because a variety of tissue respond to electrical stimulation [2]. On the other hand, an accurate and comprehensive description of structures is of vital importance for the development of high-quality bionanocomposites, which is, however, very difficult because of their nature. Electrical conductivity depends on the micro- and macroscopic states of the composites and gives information on the interaction of individual components inside. Determining the electrical properties of a material is therefore one of the most convenient and sensitive methods for studying the structures of materials [3, 4]. The dielectric properties of bionanocomposites depend on the nature of biopolymer matrix and filler particles, the dimension (mainly include length-to-diameter or aspect ratio) and content of filler particles, and the interfacial characteristics. These issues are taken into consideration in this chapter when reviewing the dielectric performance of bionanocomposites.

7.2 Dielectric Properties of Bio-Assembled Nanocomposites

Nature itself synthesizes and fabricates hybrid composites with hierarchical architectures that exhibit extraordinary properties, which can be used to make new and biological electric, photonic, and magnetic devices. Two natural composites that have calcium carbonate ($CaCO_3$) as one of the main components and present superior dielectric properties are exemplified.

Seashells are composed of well-aligned and highly packed aragonite nanolaminas glued together by biopolymer, whose superior mechanical properties have been reported [5, 6]. Conch shells are one of the examples, whose microstructures are shown in Fig. 7.1 [7].

Figure 7.1 (a) Photograph of the conch shell used with the indication of how the samples were cut. (b) SEM image showing the microarchitecture of the conch shell. (c) SEM image showing how the nanoaragonite laminas are assembled. (d) TEM image of a cross section of the nanolaminas and the bio-organic matrix between the nanolaminas. Reprinted from Ref. [7], with permission from John Wiley and Sons.

Yao et al. [7] measured the dielectric properties of the sample cut from the conch shell in the frequency range from 100 Hz to 10 MHz at room temperature. Figure 7.2 shows representative curves of the dielectric characteristics as a function of frequency measured on different samples. The relative dielectric constant (ε_r) lay in the range of 80–300 at the low frequency of 100 Hz and decreased to ~13–19 at 1 MHz. The dielectric loss (tan δ) decreased from 0.3–1.0 at 100 Hz to 0.05–0.3 at 1 MHz. These properties differ significantly from those of single-crystal slices of $CaCO_3$ with a thickness of 0.5 mm whose dielectric constant is ~9 and independent of frequency. Moreover, tan δ of a single crystal of $CaCO_3$ is very small, less than 0.01, across the entire frequency range. These comparisons indicate that the electrical properties of the bio-assembled nanocomposite differ from those of single crystals such as $CaCO_3$. The large dielectric loss of the conch implies a significant electrical leakage. Furthermore, the nanolaminas and the biopolymer layers of conch shells exhibited square polarization–electric field (*P*–*E*) hysteresis loops with giant "polarization" and pyroelectricity, which are the indication of ferroelectret behavior. These electrical behaviors originate from the charged nanocapacitors that are created by the hierarchical microarchitecture of the $CaCO_3$ nanocomposites, in which biopolymer layers form a complex 3D network. The electrical performance of such nanomaterials, such as the polarization, coercive field, and reversal frequency of the giant dipoles, can be tuned by varying the properties of the polymers and the size of the nanolaminas for varied applications utilizing ferroelectret materials, including electret motors and generators as well as of high-density energy storage.

The other typical natural nanocomposite is pearl. Natural pearls are composed of $CaCO_3$ and a tough, horny organic matter called conchiolin. The use of pearl has been limited to jewelry, cosmetic, and pharmaceutical applications. Pioneering work was carried out by Pradhan and Sarkar [8] to study the electroactive and dielectric performance of natural pearls in order to explore the potential of pearls in technological applications.

Figure 7.3 summarizes the result of DC volt–ampere characteristics of the tested pearl. The recorded discontinuities in the *V*–*I* curve implies the poor surface conductivity and probable presence of nanostructure over the pearl surface. Figure 7.4 shows the variation of dielectric constant of tested pearl with temperature and frequency, respectively, measured in impressed AC field. Figure 7.4(1) indicates the nonlinear dielectric behavior, and Fig. 7.4(2) indicates the super-dielectric nature—the dielectric constant determined following a Lorentzian fit at low frequencies was found to be $\sim 10^5$. Results shown in Figs. 7.3 and 7.4 reveal that such super dielectric behavior of pearl could not be explained using Maxwell–Wagner interfacial polarization in which a large surface electrical conduction is exhibited. Instead both bulk and surface conduction of the pear specimen exhibited the features like that of nanocomposites [9], which coincided with the previously measured surface particle dimension using atomic force microscopy (AFM). Apex sharpness in nanoscale introduces a field discontinuity and may generate an electric field of almost infinite intensity. This phenomenon is termed lightning rod effect (LRE) [10], which causes an ultrahigh polarization in the dielectric background in a nanocomposite. In this study [8] the LRE was used to explain the origin of the super-dielectric nature of pearl, since the overall nature of the peal can be considered as a nanocomposite of small-sized metallic and other nanoparticles in/on a calcinated background. The dielectric characteristics make natural pearl a promising biomaterial in developing ultra-capacitors and memory devices.

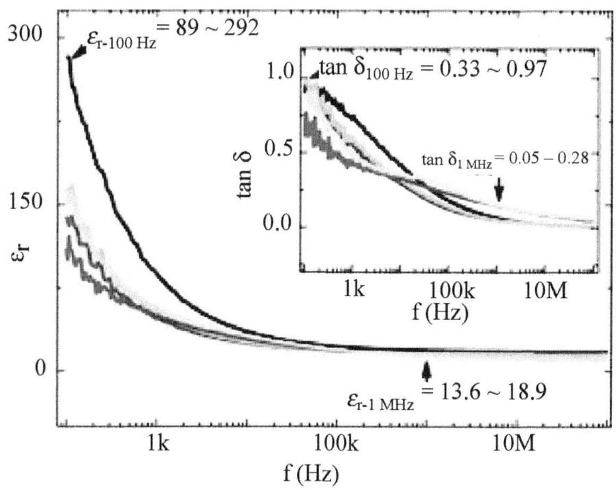

Figure 7.2 Frequency dependence of dielectric constant and loss for several representative samples from a conch shell. Reprinted from Ref. [7], with permission from John Wiley and Sons.

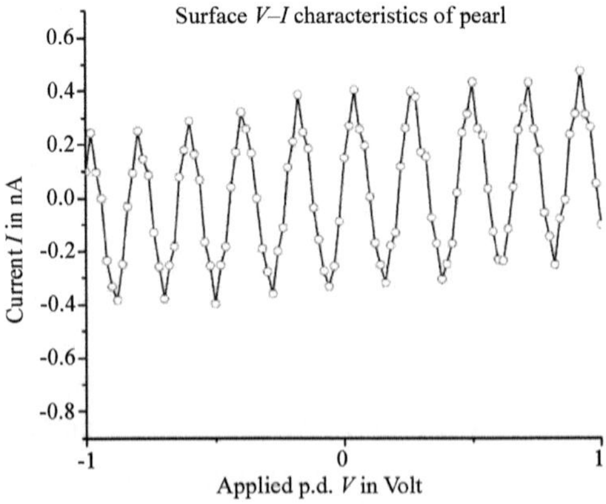

Figure 7.3 Surface DC V–I characteristics of pearl bead measured at room temperature (sample size L = 1.0 cm, CS area = 1 cm^2). Republished with permission of Trans Tech Publications, from Ref. [8], Copyright 2011; permission conveyed through Copyright Clearance Center, Inc.

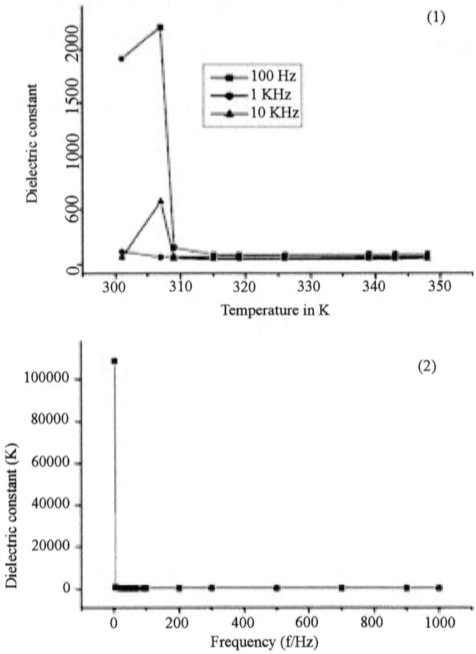

Figure 7.4 Variation of dielectric constant with (1) temperature at field frequencies of 100 Hz, 1 kHz, and 5 kHz, and (2) frequency of the impressed electric field at room temperature. Republished with permission of Trans Tech Publications, from Ref. [8], Copyright 2011; permission conveyed through Copyright Clearance Center, Inc.

7.3 Dielectric Properties of Nanocomposites with Bio-Based Components

The artificial dielectric nanocomposites with at least one component (polymer matrix or filler particles) from renewable sources or being biodegradable are reviewed in this section.

7.3.1 Biofibril-Based Nanocomposites

7.3.1.1 Natural fiber–based nanocomposites

Strictly speaking, natural fiber–reinforced polymers are not really within the scope of this chapter against nanoscale objects. However, there is currently a trend to produce multifunctional hybrid composites using filling particles ranging from nano- to macroscale, in which natural fibers and their fabrics are largely used. Therefore, the dielectric properties of natural fiber–reinforced polymer composites are still briefly reviewed.

Sreekumar et al. [11] studied the electrical properties of sisal/polyester composites at the frequencies ranging from 5 Hz to 300 kHz and temperature ranging from 25°C to 60°C. They found that all the characteristics like dielectric constant, conductivity, dissipation factor, and loss factor increased with fiber content for the entire range of frequencies. Kiew et al. [12] compared the dielectric properties of chicken feather fiber (CFF) or kenaf fiber (FF)-reinforced unsaturated polyester composites. The dielectric constant, loss factor, and dissipation factor of the composites were found to increase with either CFF or KF. The increase in these characteristics was more apparent with a fiber content of 40 vol%, at which the orientation and interfacial polarization were manifested to a larger extent. Paul and Thomas [13] reported that the electrical conductivity of hydrophobic low-density polyethylene (LDPE) can be improved by mixing it with hydrophilic lignocellulosic fibers and conductive carbon black. Hong and Wool [14] developed a bio-based composite from soybean oil and keratin feather fiber (KFF) suitable for electronic applications. It was found that the hollow KFFs were not filled by resin infusion and the composite retained a significant volume of air in the hollow structure of the fibers. Because of the retained air the dielectric constant of the composite was in the range of 1.7–2.7 depending on the fiber volume faction, and these values are significantly lower than the conventional silicon dioxide or epoxy, or polymer dielectric insulators.

Kumar et al. studied the dielectric properties of bamboo fiber–reinforced epoxy or polylactic acid (PLA) composites incorporated with nanoclay in order to obtain relatively low dielectric constant nanodielectrics with improved mechanical properties [15, 16]. It was found that silane coupling agents and mercerization of bamboo mats resulted in decreased dielectric constant and dissipation factor of the nanocomposites due to the reduction in orientation polarization by the decreased hydrophilicity compared to untreated fiber composites [15]. Nanoclay was found to improve the compatibility between the bamboo fiber and PLA matrix. The addition of 3 wt% nanoclay led to lower dielectric constant and dissipation factor of the nanocomposites compared to virgin bamboo/PLA composites, due to immobility of PLA chains by nanoclay [16].

7.3.2 Cellulose Nanocrystal–Based Nanocomposites

Since the first time Rånby [17] reported obtaining colloidal suspensions of cellulose (cellulose nanocrystals) by sulfuric acid hydrolysis of cellulose fibers, cellulose nanocrystals (CNCs) have been extracted from various biomasses such as cotton [18], ramie [19], flax [20], sisal [21], sugar beet [22], wheat straw [23], palm [24], softwood [25] and hardwood [26] pulp, microcrystalline cellulose (MCC) [27], bacterial cellulose [28], and tunicate [29]. The extracted nanocrystals are usually in a rod shape having a typical dimensions ranging from 5 to 10 nm in diameter and from 80 to 200 nm in length [30], although these geometrical characteristics depend on the nature of cellulose source and the hydrolysis conditions to a certain extent, and are therefore given the common name cellulose nanowhiskers (CNWs). CNW from algal cellulose (valonia, a sea plant) was reported to have a width of 20 nm and length of 1000–2000 nm, and that from tunicate (a sea animal) gives a width of 10–20 nm and length of 500–2000 nm. The aspect ratio (length/diameter ratio) of CNWs varies between ca. 10 and 30 for cotton and ca. 70 for tunicate. The Young's modulus of CNWs estimated by different researchers ranges between 130 GPa31 and 250 GPa [32], close to the modulus of the perfect crystal of native cellulose. The experimental strength was assessed to be of the order of 10 GPa [33]. The renewability, global presence, huge specific surface area, and charged surface originating from the way they are extracted, high aspect ratio, and impressive mechanical properties make CNW a very promising reinforcement for a wide variety of polymeric matrices.

CNWs can form a percolation network within the polymer matrix introduced by hydrogen bonding between neighboring whiskers, which have been confirmed by electrical measurement [34]. The large number of hydroxyl groups on the surface of CNWs can be used to increase the polarity of the composite systems in which polymers with low polarity are used as matrices and thus enhance the interfacial polarization and permittivity. Some polymers themselves have many polar groups that may form a large number of hydrogen bonds within the molecules, limiting the mobility of the polarized groups. The numerous hydroxyl groups of CNWs can disrupt the hydrogen bonds within the molecules of these polymers and thereby further increase the polarizability of the polymers by forming new, stronger hydrogen bonds.

Ten et al. [35] prepared poly(3-hydroxybutyrate-co-3-hydroxyvalerate)(PHBV)/CNW nanocomposites using solvent exchange followed by solution casting and studied the effect of the content and aspect ratio of CNWs isolated from MCC on the dielectric properties of the resultant composites over 0.01–10^6 Hz frequency range at room temperature. It is noteworthy that PHBV is one of the most studied environment-friendly poly(hydroxy alkanoates) (PHA) resins, whose applications are limited by some of its inherent drawbacks such as high brittleness and relatively difficult processing due to the low melt viscosity and narrow processing window. The real permittivity and real conductivity of neat PHBV and PHBV/CMW composites are shown in Fig. 7.5. The dielectric constant (ε') was found to peak at 2.3 wt% CNWs in the studied frequency range, and decreased for all CNW concentrations with increasing frequency (Fig. 7.5(1)). The specific conductivity of the composite, by contrast, increased monotonically with increasing CNW content (Fig. 7.5 (2)). Because of the different dielectric constants and conductivity between PHBV and CNW, interfacial polarization occurred, which contributed to the variation of permittivity. PHBV molecules possesses low polarity and, therefore, show lower permittivity compared with the composite because of the presence of polar

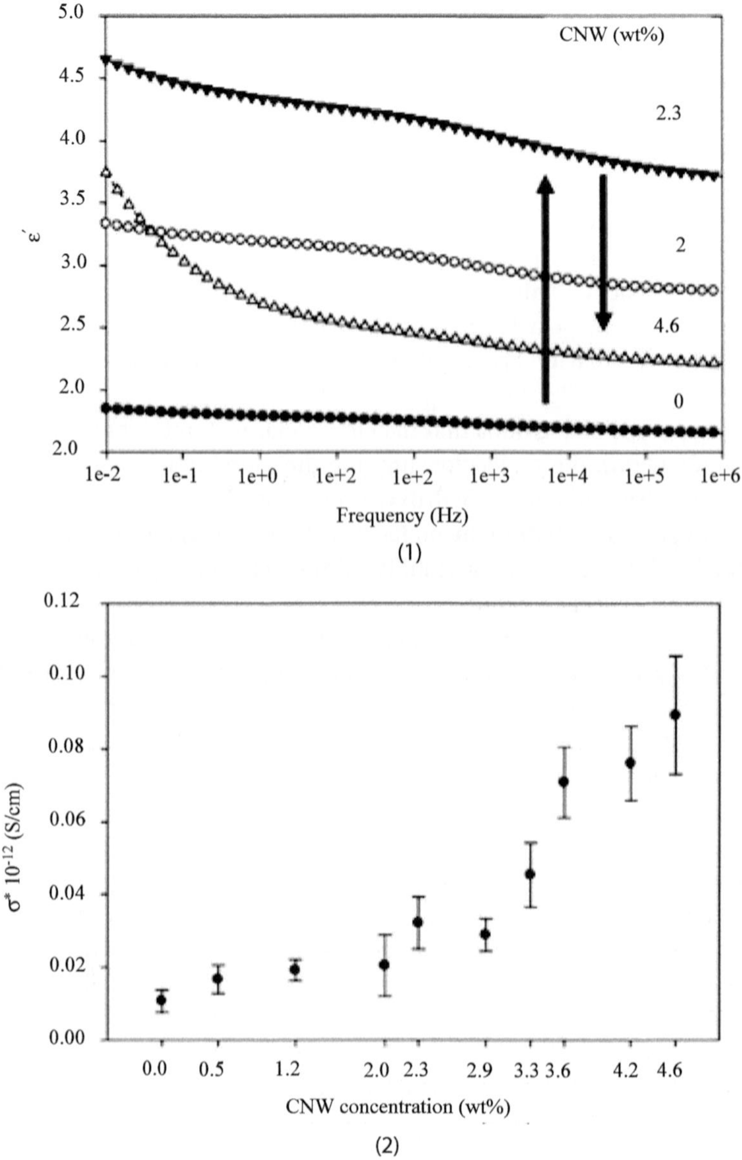

Figure 7.5 Dielectric properties of PHBV and PHBV/CNW nanocomposites: (1) real permittivity at room temperature in electric fields of different frequencies, and (2) real conductivity at frequency of 1 Hz at room temperature. Reprinted with permission from Ref. [35], Copyright 2012, American Chemical Society.

groups such as hydroxyls and sulfates on CNWs. The interfacial area between PHBV and CNWs and the interconnections between CNWs increased with CNW concentration, resulting in increased ε' up to 2.3 wt%. Above this concentration, ε' started to decrease because of the reduced number of interfaces and interconnections due to deteriorated dispersion or more agglomeration of CNWs. In addition, high conductivity of the composites at high CNW contents could also reduce the permittivity. The decreased permittivity with increasing frequency for all CNW concentrations could be attributed to the relatively slow rotation and motion of the dipoles which was not able to follow the applied electric field at high frequencies.

Tian et al. [36] prepared gelatin/glycerol (GG) dielectric elastomer (DE) with a mass ratio of 1:0.5 aiming for the development of dielectric elastomer actuators (DEAs), which possesses high dielectric constant and moderate elastic modulus. DEs have advantages, including large deformation, mechanical compliancy, light weight, quiet operation, and low cost, over other electroactive polymers (EAPs). Both gelatin and glycerol have a large number of polar groups. The hydroxyl groups in glycerol can disrupt the hydrogen bonds of C=O···H–N in gelatin molecules. Continuous work in this group was carried out using CNWs to improve the mechanical strength and further largely increase the dielectric constant of the GG elastomer [37]. They found that the conductivity of plain GG (0.66×10^{-8} S/m) slightly increased with CNW content (1.32×10^{-8} S/m at 5 wt% CNW), although both GG and CNW are electrically insulated. It was ascribed to the further disruption of hydrogen bonds in gelatin by CNWs. The variation of electromechanical properties of GG with increasing CNW content (Table 7.1) shows that (i) the dielectric constant at the same frequency of GG increased with increasing content of CNWs, also revealing the disruption of hydrogen bonds between gelation molecules by hydroxyl groups on CNWs; (ii) the elastic modulus (Y) of GG increased with increasing CNW content, ascribed to the good interfacial adhesion between CNWs and GG, leading to a good dispersion of CNWs within the matrix blend; (iii) the relaxation peak of dielectric loss for plain GG shifted to higher frequency for GG/CNW composites, ascribed to the good interfacial interaction between CNWs and GG matrix; (iv) electromechanical sensitivity (β), defined as the ratio of the dielectric constant (ε_r) to the elastic modulus (Y), which is required to be high for achieving a DE with high actuated strain at a low electric field, increased as CNW content increased; (v) the actuated strain at the same electric field increased with increasing CNW content, ascribed to the increase in the dipole rotation and/or the polarizability of GG caused by the further disruption of hydrogen bonds. The incorporation of CNW is therefore considered as an efficient method to prepare high-performance bio-based DE for applications in biological and medical fields [37].

Table 7.1 Mechanical properties and electromechanical properties of CNWs/GG composite films with different contents of CNWs measured at room temperature

Filler content (wt%)	Dielectric constant ε_r at 10^2 Hz	Dielectric loss tangent at 10^2 Hz	Elastic modulus Y (MPa)	$B = \varepsilon/Y$ (MPa^{-1})	Actuated strain at 0.3 kV/mm (%)
0	520	1.7	1.4	373	0.30
1	630	2.0	1.6	392	0.36
2.5	810	2.1	1.9	421	0.57
5	930	2.3	2.1	462	0.99

Source: Ref. [37]

The promising applications of cellulose expand to electronic, sensor, and display fields; however, cellulose generally has a passive function by acting as substrate instead of an active role in the device's working principle [38]. Gaspar et al. [39], within EU projects INVISIBLE (FP7 ERC AdG No. 228144), ORAMA (FP7 CP-IP 246334-2), and APPLE (FP7-NMP-2010-SME/262782-2), initiated a thin transparent CNW paper-based field-effect transistor (FET) where CNW was simultaneously used as the substrate and the gate dielectric layer in an "interstrate" structure, since the device was built on both sides of the CNW films. The active channel layer was based on oxide amorphous semiconductors and the gate electrode was based on a transparent conductive oxide. The device configuration including all the fabrication steps as well as the different layers and corresponding thickness is shown in Fig. 7.6. The transistors processed in this way had an enhancement of the n-type operation mode and exhibited a channel saturation mobility exceeding 7 cm^2

$V^{-1} s^{-1}$, a drain–source current I_{ON}/I_{OFF} modulation ratio above 10^5 and a subthreshold gate voltage swing of about 2 V/decade. They also presented good stability after two weeks of being exposed in air ambient conditions, without any type of encapsulation or passivation layer. The results offered a promising approach for attaining high-performance disposable electronics such as paper displays, smart labels, smart packaging, radio-frequency identification (RFID), and point-of-care systems for self-analysis in bioscience applications.

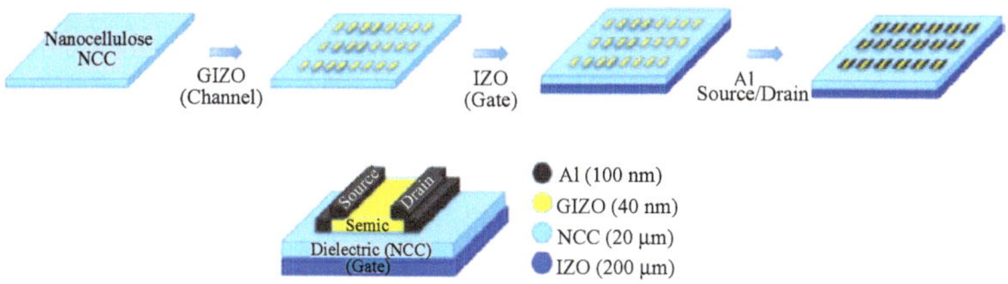

Figure 7.6 Schematics of the process of fabrication of FETs using CNW as the gate dielectric, and the corresponding staggered-bottom gate structure used in Ref. [39], where Al refers to aluminum, GIZO refers to Ga_2O_3-In_2O_3-ZnO (1:2:2 mol%), NCC refers to CNW, and IZO refers to In_2O_3-ZnO (5:2 mol%). Reproduced with permission from Ref. [39], Copyright 2014, IOP Publishing. All rights reserved.

7.3.3 Carbon Nanomaterial–Based Nanocomposites

Carbon nanostructures include fullerene (buckyballs), carbon nanotubes (single-walled, double-walled, and multiwalled), carbon nanofibers, carbon nanoparticles, and graphene nanosheets.

An exponential growth on polymer nanocomposite reinforced with carbon nanotubes (CNTs) occurred after the first research work published by Ajayan et al. [40]. Besides the extremely high theoretical/experimental tensile strength (150–180 GPa) and modulus (640 GPa to 1 TPa) [41], the one-dimensional electronic structures of CNTs make them the most appealing class among conductive nanomaterials. Such unique electrical properties have been exploited to modify the dielectric properties of bio-derived, biocompatible, and biodegradable materials for a range of technological and bioelectronic applications.

Dionigi et al. [42] produced a conductive silk fibroin (SF)-single-walled carbon nanotube (SWCNT) film using the templating method, which had 3D ordered structures of interconnected submicrometric pores, as shown in Fig. 7.7. SWCNTs were assembled in periodic structures by using polystyrene beads (PB), acting as both carriers and a sacrificial template, which were then infiltrated by silk fibroin. This method was considered versatile, which allowed the tuning of the periodicity of the SF-SWCNT porous structure by controlling the templating bead diameter. The nanostructured SF-SWCNT composite showed a conductivity (δ at 0.3 V is 2.81 S cm^{-1}) just one order of magnitude lower than that of the bare SWCNTs (δ at 0.3 V is 75.25 S cm^{-1}). It indicates that the conductivity functionality has been added to the bare SF film, which was originally an electrical insulator. Furthermore, the drop-casting SF onto the SWCNTs did not destroy the SWCNT interconnections. The results are expected to greatly facilitate the fabrication and development of silk-based electronic devices for use in bioelectronic and biomedical applications.

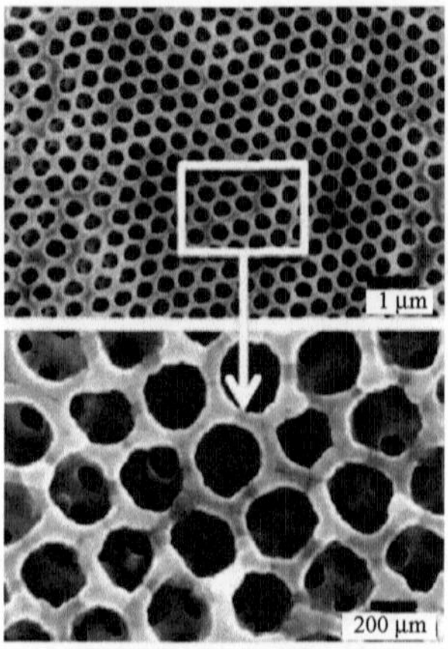

Figure 7.7 SEM micrographs of a nanostructured silk fibroin-SWCNT porous film at different magnifications. Reprinted from Ref. [42], with permission from Royal Society of Chemistry.

Chitosan (CTS)/CNT represents a very attractive bionanocomposite whose preparation, structural characterization, and mechanical properties, and application in biosensors [43–46] have been studied. Perez et al. [47] used dynamic mechanical analysis and dielectric spectroscopy to study the relaxation phenomena of CTS/MWCNT bionanocomposites. It was considered as the first study on the dielectric properties of such bionanocomposites. Hernández-Vargas et al. [48] further studied the effect of silver nanoparticles (AgnP) decorated CNT on the dielectric properties of CTS measured at a 100 Hz to 110 MHz frequency range and a 2–250°C temperature range. The conductivity of pristine CTS was enhanced by six orders of magnitude with the inclusion of MWCNTs, while the addition of CNT-AgnP only increased two orders of magnitude. It was, as revealed by FTIR analysis, because of strong chemical interactions between CNT and AgnP. The percolation threshold in the nanocomposites therefore increased from ~1.3 wt% when using MWCNTs to ~2.2 wt% when using CNT-AgnP. Irrespective of the decoration of AgnP, DC conductivity of CTS/CNT exhibited a 3D hopping conductivity, the value of the critical exponent is ~2.08 and 1.87 for CTS/CNT and CTS/CNT-AgnP, respectively. In contrast, when CTS was combined with only AgnP the composite exhibited a 2D hopping conductivity with a critical exponent of ~1.32, indicating that the presence of CNT in composites led to 3D conductivity systems.

Aspect ratio and content of CNTs are the two key factors affecting the dielectric properties of CNT-based composites. Musto et al. [49] studied the electric properties of a nanocomposite system composed of PHBV/poly(butylenes adipate-co-terephthalate)(PBAT) blend and MWCNTs processed via melt compounding. It was found that the dielectric constant increased remarkably with MWCNT content especially at low frequencies, which was explained in terms of a gradual formation of micro-capacitor networks [50, 51] where many MWCNTs agglomerates are separated by very thin dielectric polymer layers. While at low frequencies (i.e., 1 kHz, as shown in Fig. 7.8) and at low MWCNT content far from the percolation threshold, the relative permittivity of the composites remained practically unchanged, almost equal to that of the neat resin (ε_r = 3.8). With increasing MWCNTs content the number of micro-capacitors increased and the insulation distance between particles decreased, so that the capacitance of a single micro-capacitor increased. From Fig. 7.8 it was also seen that in the composite with high-aspect ratio (667) MWCNTs the relative permittivity sensibly increased above the percolation threshold φ_c = 0.42%, while in the composite with medium aspect ratio (105) the threshold was not the percolation threshold but the transition threshold φ_t = 0.06%, above which relative permittivity changed linearly but with a much lower speed compared to the high-aspect ratio counterpart. The dissipation factor (tan δ) (Fig. 7.9) of the neat PHBV/PBAT blend was extremely low in all the measured frequency range, which increased with frequency though and assumed its maximum (0.06) at frequency of 10 MHz. In composites with low–aspect ratio (30) MWCNTs, the variation of tan δ with CNT content was almost negligible. In composites with medium–aspect ratio (105) MWCNTs tan δ remained practically unchanged at different CNT contents and increased only above the percolation threshold. In composites with high–aspect ratio (667) MWCNTs, tan δ was more than 8 at low frequency and high CNT content.

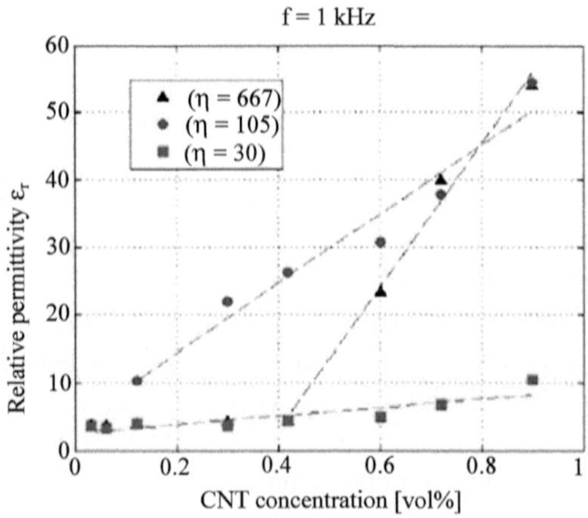

Figure 7.8 Relative permittivity at 1 kHz vs. MWCNT content at different aspect ratios. Reprinted from Ref. [49], Copyright 2015, with permission from Elsevier.

Graphene was also found able to expand the applications of bio-based polymers as capacitor. Sahila and Jayakumari [52] prepared graphene/benzoxazine-epoxy nanocomposite by first using a bio-product to synthesize new benzoxazine monomer, then in situ polymerizing benzoxazine–graphene nanocomposites (pba-g), followed by copolymerization with diglycidyl ether of bisphenol A at a loading from 20% to 60% using pba-g precursor. The dielectric characteristics can be seen in Fig. 7.10. The dielectric constant increased with increasing content of pba-g over the entire frequency range. At 10^4 Hz the dielectric constant of 40 pba-g was 60 times higher than that of neat epoxy resin. This was attributed to the motion of free charge carriers due to the formation of continuous conductive pathway throughout the nanocomposite. The capacitance value also increased with the increasing level of pba-g to the epoxy, due to the existence of strong interfacial polarization between the filler and the blend, clearly indicating that the percolation threshold has been reached in the pba-g nanocomposite with a weight fraction of 20 pba-g. The dielectric loss factor of 60 pba-g was not only dependent on the filler content but also on frequency, while for 20 pba-g and 40 pba-g it was independent of frequency over a wide range.

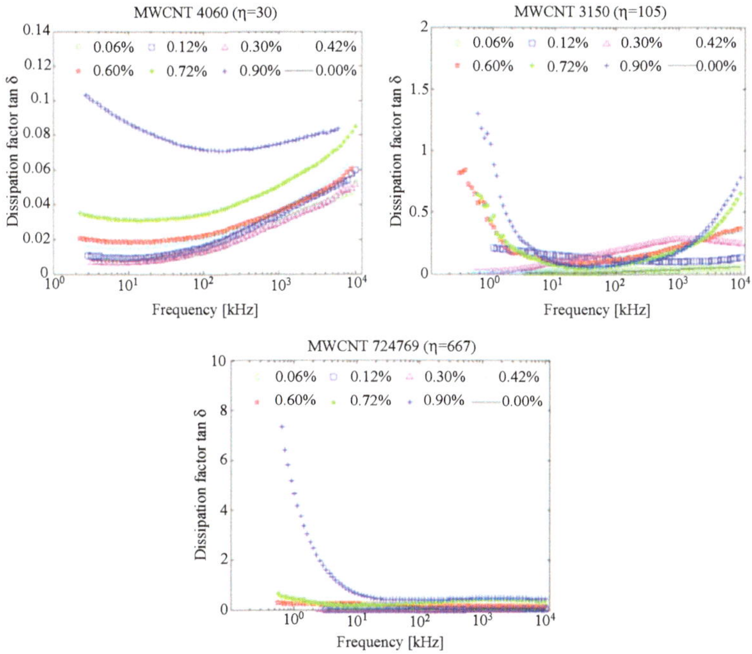

Figure 7.9 Dissipation factor vs. frequency at different MWCNT content. Reprinted from Ref. [49], Copyright 2015, with permission from Elsevier.

7.3.4 Ceramic Oxide–Based Nanocomposites

Ceramic oxides [53] represent the most extensive group of ceramic materials produced today. Traditionally, but rather artificially, the oxide ceramics are divided into "traditional" and "advanced" groups. The "traditional" ceramics include mostly silica-based products prepared from natural raw materials (clays), including building parts (bricks, tiles), pottery, sanitary ware, and porcelain, but also ceramics with other main components (e.g., alumina, magnesia), which are applied in the field of electroceramics (insulators), or industrial refractories. "Advanced" ceramics require a much higher quality and purity of raw materials, as well as the careful control of processing conditions and of the materials' micro-structure. They usually comprise oxides, which do not quite fall within the traditional understanding of the term *silicate* materials and ceramics. Oxides found in these ceramics include mostly oxides of metals such as aluminum, zirconium, titanium, and rare earth elements. In this section typical ceramics, both "traditional" and "advanced" that were incorporated into bio- and bio-based polymers to change/modify the dielectric properties of the matrices for their bio- and energy storage applications, will be briefly introduced.

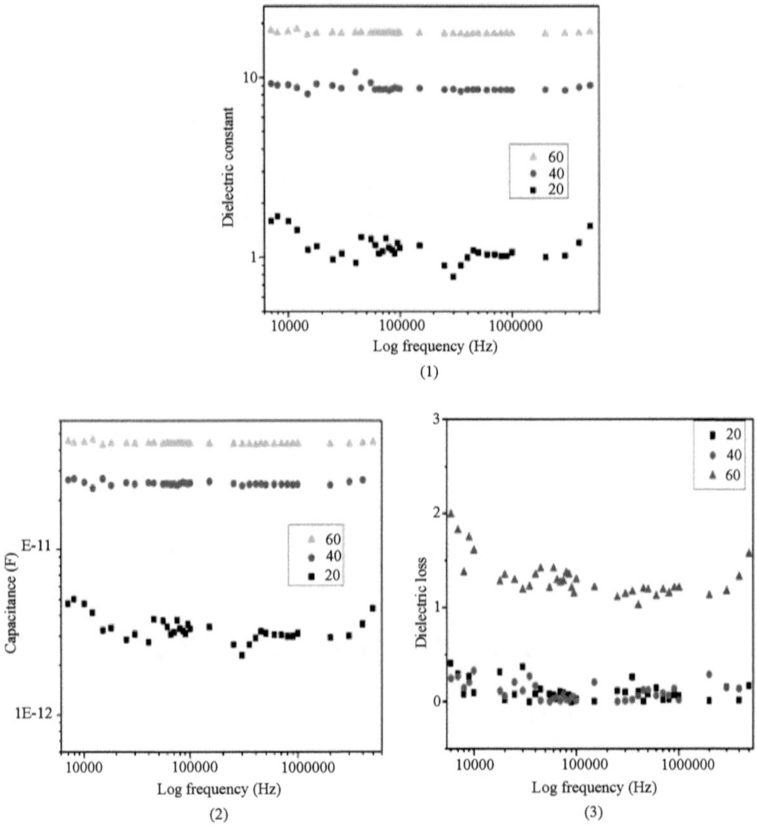

Figure 7.10 Dielectric characteristics of graphene/benzoxazine-epoxy nanocomposite with 20 pba-g to 60 pba-g: (1) dielectric constant, (2) capacitance, and (3) dielectric loss. Reprinted from Ref. [52], with permission from John Wiley and Sons.

7.3.4.1 "Traditional" ceramic–based nanocomposites

A surface functionalized 2D mesoporous silica (F-SBA-15) was obtained by extraction from rice husk ash followed by functionalization using 3-glycidoxypropyl trimethoxy silane. It was then incorporated into epoxy resin with varying contents 1, 3, and 5 wt%, for the development of high-performance nanocomposites for low dielectric constant microelectronic applications [54]. The variation of dielectric constant of the nanocomposites in frequency range between 100 Hz to 1 MHz at 30°C is shown in Fig. 7.11. The values of dielectric constant were reduced from 4.48 for the neat epoxy to 2.73, 2.44, and 2.14 for 1, 3, and 5 wt% F-SBA-15 reinforced nanocomposites. It was ascribed to the presence of more number of air voids (whose dielectric constant is unity) resulted from mesoporous structure of silica, which interrupted the interaction between the matrix molecules, and from enhanced free volume of the resulting composites. In addition, the value of dielectric constant of the nanocomposites was also dependent upon the frequency. This value decreased with increasing frequency due to polarizable mechanism since the response of electronic, atomic, and dipolar polarizable units varies with frequency.

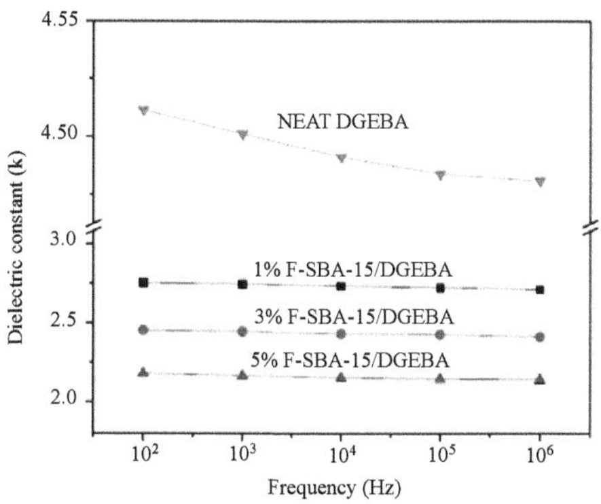

Figure 7.11 Dielectric constant profile of bio-based F-SBA-15 reinforced epoxy nanocomposites. Reprinted with permission from Ref. [54], Copyright 2014, SAGE Publications, Ltd.

Halloysite nanotubes (HNTs) are two-layered aluminosilicate minerals, available in abundance in many locations around the world. It is chemically similar to kaolin differing mainly in the morphology of crystals [55]. They possess hollow nanotubular structure with length of 1–15 μm and inner and outer diameter of 15–100 and 40–120 nm, respectively. HNTs are naturally available, biocompatible, and much cheaper compared to CNTs. Moreover, HNTs have low density of hydroxyl groups when compared to virgin clays, which makes them disperse in a polymer matrix better than clays and/or CNTs. Highly dispersed HNTs was seen to bring favorable changes in the thermal stability, dielectric, and rheological characteristics in some engineering polymers, such as polyamide-11 (PA-11) [56]. It was found that melt extrusion using a masterbatch dilution process led to a highly dispersed state of HNTs in PA-11 matrix with content of 2, 4, and 6 wt%. The dielectric constant of neat PA-11 increased with increasing HNTs content over the whole studied frequency range between 50 Hz and 1 MHz. It was attributed to the polar characteristics of HNTs and thus increased number of charge carriers in the PA-11 matrix. Such dielectric performance was consistent with a previous study on CNT/PP nanocomposite [57].

Bio-ceramic nano-hydroxyapatite (NHAp) ($Ca_5(PO_4)_3OH$) is chemically similar to the mineral components of bones and hard tissues in mammals and finds applications in membrane separation, dental applications, chromatography, gas sensors, and high-temperature fuel cells [58]. It has been incorporated, with a varied content of 5, 10, and 15 wt%, into a polymeric electrolyte film consisting of poly (methyl methacrylate) (PMMA) (80–95 wt%) as matrix and lithium bisoxalatoborate (LiBOB) (5 wt%) as doping salt to enhance the electrochemical properties and stiffness [59]. The addition of NHAp significantly enhanced the ionic conductivity and blocking the reorganizing tendency of polymer matrix by possibly acting as crosslinking centers for PMMA segments. The maximum ionic conductivity was found to be in the range of $10^{-4.8}$ S/cm for the nanocomposite containing 10 wt% NHAp.

7.3.4.2 "Advanced" ceramics–based nanocomposites

Metal oxides which exhibit stability under harsh processing conditions and safety in animal and human applications, moreover, offer antimicrobial, antifungal, antistatic, and UV-blocking properties [60], are used extensively to construct functional coatings and bionanocomposites [61–63]. ZnO [64, 65] and TiO_2 [66] are two typical metal oxides for such applications. They are both multifunctional n-type semiconductors, besides being nontoxic and environment-friendly.

ZnO nanorods (ZnO-NRs) have been incorporated into gelatin [67] and starch [68] to improve the mechanical, electrical, and optical properties of the biopolymers. The conductivity properties of the gelatin/ZnO-NR and starch/ZnO-NR nanocomposites were examined in the frequency range of 0.01 kHz and 1 MHz. It was found that the conductivity of both neat polymers was significantly increased with the addition of ZnO-NRs, which further increased with increasing particle contents. The maximum conductivity of 0.92×10^{-6} S cm^{-1} was observed for the gelatin/ZnO-NR nanocomposite with 5 wt% ZnO-NRs. The electrical conductivity and relative dielectric constant of starch was increased by 53% (to 0.95 μS cm^{-1}) and 60% (to 44.1), respectively, by 10 wt% ZnO-NRs. The highest obtainable room temperature conductivity was 0.95×10^{-6} S cm^{-1} for the starch/ZnO-NR nanocomposite with 10 wt% ZnO-NRs.

TiO_2 nanoparticles has been incorporated into a bio-based blend of poly(ether ether ketone)(PEEK)/poly(ether imide)(PEI) via ultrasonication followed by melt-blending [69] for enhancing the load-bearing capability of PEEK which is a multipurpose biomaterial greatly suitable for the development of medical device applications [70]. The dielectric constant (ε') value of PEEK was relatively low, ~3 at 10 Hz and remained almost constant at higher frequencies. PEEK/PEI blend showed similar behavior, just a slightly higher value. The addition of polar TiO_2 caused a noticeable increase in ε' of the blend particularly at low frequencies (below 1×10^3 Hz). Almost 2-fold increase was attained at the highest nanoparticle loading at 10 Hz, attributed to the high ε' of TiO_2, which improved the polarization via dipole–dipole interactions among neighboring nanoparticles. Furthermore, the nanocomposites showed a marked increase in ε' with decreasing frequency. This was clearly related to the presence of the nanoparticles and suggested a Maxwell–Wagner interfacial polarization. The dielectric loss (ε'') also rose gradually with increasing TiO_2 loading, which was also related to the increase in the space charge density. These results indicated that the nanocomposites with inclusion of TiO_2 are better dielectric materials for the development of tissue scaffolds than the neat polymers of PEEK. TiO_2 was also used to modify the dielectric properties of DNA biopolymer film for energy storage applications [71]. The dielectric constant of DNA was considerably enhanced in the entire frequency range of 1 kHz to 1 MHz at room temperature. The dielectric constant of neat DNA at RT was 7.9 at 1 kHz and 6.8 at 1 MHz, while it was increased to 9.0 at 1 kHz and 8.0 at 1 MHz with 20 wt% TiO_2, and 14.3 at 1 kHz and 11 at 1 MHz with 40 wt% TiO_2. The bulk resistivity of DNA-TiO_2 film was measured to be two to three orders of magnitude higher than that of the neat DNA film. These results indicated the potential of the DNA-TiO_2 film for utilization as insulating dielectrics in capacitor applications. The variable temperature dielectric measurements in the range of 20–80°C range revealed that TiO_2 stabilized DNA against large temperature-dependent variations in both dielectric constant and dielectric loss factor.

7.3.5 Precious Metal–Based Nanocomposites

Gold nanoparticles are among the most stable metal nanoparticles with some unique features and properties such as size-related electronic, optical, and magnetic properties as well as application in catalysis and biological system [72]. They have been used in recent years in core–shell structures which exhibit unique size-dependent electrical and optical properties [73].

Bacteria cellulose nanofibers with directly attached gold nanoparticles were prepared by Zhang et al. [74] and showed interesting properties for biosensing applications. Gruber et al. [75] presented for the first time the preparation of cellulose-biotemplated silica nanowires densely coated with gold nanoparticles. The nanocrystalline cellulose template was successfully coated with silica and subsequently seeded with gold nanoparticles with diameters of about 2–6 nm. They found that the optical properties could be tuned by the gold coating thickness. Variation in dielectric properties over the dielectric nanocrystalline cellulose core can also be foreseen since two localized surface plasmon resonance (LSPR) will be generated at the core–shell and shell–matrix interfaces.

Massaro et al. [76] prepared a polydimethylsiloxane (PDMS)–gold nanocomposite to wrap outside of a tapered optical fiber, which was developed into optical pressure sensors suitable for robot tactile sensing (see the illustration in Fig. 7.12). PDMS belongs to a group of polymeric organosilicon compounds, which is optically clear, and in general, inert, non-toxic, and non-flammable. It is

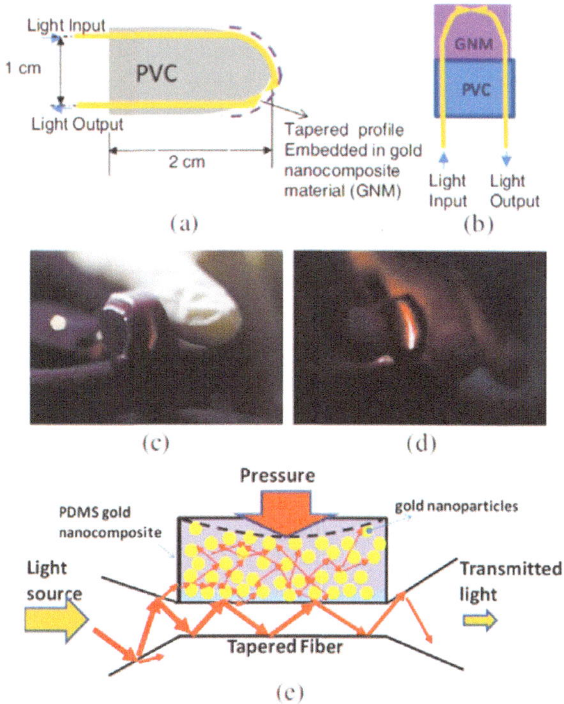

Figure 7.12 (a, b) Layouts of optical pressure sensor implemented in robotic system: a tapered optical fiber is bended on a PVC support and embedded in a PDMS-Au material (GNM). (c, d) Photos of the first prototype (the PDMS-Au cladding is 5 mm thick). (e) Light coupling and scattering process inside the GNM. Reprinted from Ref. [76], Copyright 2011, EMW Publishing.

present in shampoos, food, and caulking. It was chosen for the proposed sensor due to its ability to generate gold nanoparticles staring from gold precursor. The gold nanoparticles formed in PDMS were expected to increase the effective refractive index of PDMS and support the electromagnetic coupling with the tapered region of the fiber. The pressure applied on the nanocomposite introduced a displacement of the nanoparticles along its interface with the tapered fiber increasing the light scattering. The effect of the nanoparticle displacements due to the applied pressure is to change the effective refractive index of the nanocomposite as a function of gold concentration. Assuming spherical gold nanoparticles in PDMS, the effective dielectric function ε_{eff} for the particles having dielectric function ε_m which varied with the optical working wavelengths [77], embedded in a medium ε_s was defined as [78, 79]:

$$\varepsilon_{eff} = \varepsilon_s \frac{\varepsilon_m(1+2\phi)+2\varepsilon_s(1-\phi)}{\varepsilon_m(1-\phi)+\varepsilon_s(2-\phi)} \tag{7.1}$$

where ϕ indicated the gold concentration. The modes of the tapered fiber will exchange the power with PDMA-Au cladding by defining the coupling coefficient as [80]:

$$C_{\psi_{nm},E^r}(z) = \omega\int_{-\infty}^{\infty}\int_{-\infty}^{\infty} \Delta\varepsilon_{eff}(z)\psi_{nm}E^r dxdy \tag{7.2}$$

where ψ_{nm} are the fiber modes, E^r is the electric evanescent field radiated by the tapered profile, and $\Delta\varepsilon_{eff}$ indicated the variation of the effective permittivity of the cladding due to different applied forces.

7.4 Summary

In this chapter, the dielectric behavior and conductivity of bionanocomposites are discussed primarily in terms of the assembly structures, the species of components and their compositions, the dimensions and structural properties of the components. Novel or advanced templates for designing new (nano)structured biocomposites for biomedical, optical, and energy storage applications are also introduced.

References

1. Dufresne, A., Thomas, S., and Pothan, L. A. (2013). Bionanocomposites: State of the art, challenges, and opportunities, In *Biopolymer Nanocomposites: Processing, Properties and Applications*, Eds. Dufresne, A., Thomas, S., and Pothan, L. A., pp. 1–10, John Wiley & Sons.
2. Hardy, J. G., Lee, J. Y., and Schmidt, C. E. (2013). Biomimetic conducting polymer-based tissue scaffolds. *Curr Opin Biotechnol*, **24**, 847–854.
3. Basavaraja, C., Kim, N. R., Jo, E. A., and Huh, D. S. (2010). Biological templating of polyaniline and polypyrrole using *E coli*. *Macromol Res*, **18**, 222–226.

4. Hernández-Vargas, J., González-Campos, J. B., Lara-Romero, J., Prokhorov, E., Luna-Bárcenas, G., Aviña-Verduzco, J. A., and González-Hernández, J. C. (2014). Chitosan /WMCNTs-decorated with silver nanoparticle composites: dielectric and antibacterial characterization. *J Appl Polym Sci,* **131**, 40214.
5. Smith, B. L., Schaffer, T. E., Viani, M., Thompson, J. B., Frederick, N. A., Kindt, J., Belcher, A., Stucky, G. D., Morese, D. E., and Hansma, P. K. (1999). Molecular mechanistic origin of the toughness of natural adhesives, fibers and composites. *Nature,* **399**, 761–763.
6. Tang, Z. Y., Kotiv, N. A., Magonov, S., and Ozturk, B. (2003). Nanostructured artificial nacre. *Nat Mater,* **2**, 413–418.
7. Yao, Y., Wang, Q., Wang, H., Zhang, B., Zhao, C., Wang, Z., Xu, Z., Wu, Y., Huang, W., Qian, P. Y., and Zhang, X. X. (2013). Bio-assembled nanocomposites in conch shells exhibit giant electret hysteresis. *Adv Mater,* **25**, 711–718.
8. Pradhan, S. S. and Sarkar, A. (2011). Pearl – a nano-composite & natural super dielectric. *J Biomim Biomater Tissue Eng,* **11**, 1–12.
9. Gupta, N., Mallik, H., and Sarkar, A. (2005). Electrical conductivity of quantum size 2d nanocomplex. *J Metastable Nanocryst Mater,* **23**, 335–338.
10. Gopinath, A., Boriskina, S. V., Reinhard, B. M., and Negro, L. D. (2009). Deterministic aperiodic arrays of metal nanoparticles for surface-enhanced Raman scattering. *Opt Express,* **17**, 3731–3753.
11. Sreekumar, P. A., Saiter, J. M., Joseph, K., Unnikrishnan, G., and Thomas, S. (2012). Electrical properties of short sisal fiber reinforced polyester composites fabricated by resin transfer molding. *Compos Part A: Appl Sci Manuf,* **43**, 507–511.
12. Kiew, K. S., Hamdan, S., and Rahman, Md. R. (2013). Comparative study of dielectric properties of chicken feather/kenaf fiber reinforced unsaturated polyester composites. *Bioresources,* **8**, 1591–1603.
13. Paul, A., Joseph, K., and Thomas, S. (1997). Effect of surface treatments on the electrical properties of low-density polyethylene composites reinforced with short sisal fibers. *Compos Sci Technol,* **57**, 67–79.
14. Kong, C. K. and Wool, R. P. (2005). Development of a bio-based composite material from soybean oil and keratin fibers. *J Appl Polym Sci,* **95**, 1524–1538.
15. Kumar, V. and Kumar, R. (2012). Dielectric and mechanical properties of alkali- and silane-treated bamboo-epoxy nanocomposites. *J Compos Mater,* **46**, 3089–3101.

16. Kumar, V., Sharma, N. K., and Kumar, R. (2013). Dielectric, mechanical, and thermal properties of bamboo-polylactic acid bionanocomposites. *J Reinf Plast Compos,* **32**, 42–51.

17. Rånby, B. G. (1949). Aquesous colloidal solutions of cellulose micelles. *Acta Chem Scand,* **3**, 649–650.

18. Dong, X. M., Revol, J. F., and Gray, D. G. (1998). Effect of microcrystalline preparation conditions on the formation of colloid crystals of cellulose. *Cellulose,* **5**, 19–32.

19. Habibi, Y., Foulon, L., Aguie-Beghin, V., Molinari, M., and Douillard, R. (2007). Langmuir-Blodgett films of cellulose nanocrystals: preparation and characterization. *J Colloid Interf Sci,* **316**, 388–397.

20. Cao, X., Dong, H., and Li, C. M. (2007). New nanocomposite materials reinforced with flax cellulose nanocrystals in waterborne polyurethane. *Biomacromolecules,* **8**, 899–904.

21. Garcia de Rodriguez, N. L., Thielemans, W., and Dufresne, A. (2006). Sisal cellulose whiskers reinforced polyvinyl acetate nanocomposites. *Cellulose,* **13**, 261–270.

22. Azizi Samir, M. A. S., Alloin, F., Paillet, M., and Dufresne, A. (2004). Tangling effect in fibrillated cellulose reinforced nanocomposites. *Macromolecules,* **37**, 4313–4316.

23. Helbert, W., Cavaille, J. Y., and Dufresne, A. (1996). Thermoplastic nanocomposites filled with wheat straw cellulose whiskers. Part I. processing and mechanical behavior. *Polym Compos,* **17**, 604–611.

24. Bendahou, A., Habibi, Y., Kaddami, H., and Dufresne, A. (2009). Physico-chemical characterization of palm from Phoenix Dactylifera-L, preparation of cellulose whiskers and natural rubber-based nanocomposites. *J Biobased Mater Bioenergy,* **3**, 81–90.

25. Araki, J., Wada, M., Kuga, S., and Okano, T.(1999). Influence of surface charge on viscosity behavior of cellulose microcrystal suspension. *J Wood Sci,* **45**, 258–261.

26. Beck-Candanedo, S., Roman, M., and Gray, D. G. (2005). Effect of reaction conditions on the properties and behavior of wood cellulose nanocrystal suspensions. *Biomacromolecules,* **6**, 1048–1054.

27. Bondeson, D., Mathew, A., and Oksman, K. (2006). Optimization of the isolation of nanocrystals from microcrystalline cellulose by acid hydrolysis. *Cellulose,* **13**, 171–180.

28. Araki, J. and Kuga, S. (2001). Effect of trace electrolyte on liquid crystal type of cellulose microcrystals. *Langmuir,* **17**, 4493–4496.

29. Angles, M. N. and Dufresne, A. (2000). Plasticized starch/tunicin whiskers nanocomposites. 1. Structural analysis. *Macromolecules,* **33**, 8344–8353.

30. Boluk, Y., Zhao, L., and Incani, V. (2012). Dispersions of nanocrystalline cellulose in aqueous polymer solutions: structure formation of colloidal rods. *Langmuir,* **28**, 6114–6123.

31. Sakurada, I., Nukushina, Y., and Ito, T. (1962). Experimental determination of the elastic modulus of crystalline regions in oriented polymers. *J Polym Sci,* **57**, 651–659.

32. Zimmermann, T., Pohler, E., and Geiger, T. (2004). Cellulose fibrils for polymer reinforcement. *Adv Eng Mater,* **6**, 754–761.

33. Azizi Samir, M. A. S., Alloin, F., and Dufresne, A. (2005). Review of recent research into cellulosic whiskers, their properties and their application in nanocomposite field. *Biomacromolecules,* **6**, 612–626.

34. Flandin, L., Bidan, G., Bréchet, Y., and Cavaillé, J. Y. (2000). New nanocomposite materials made of an insulating matrix and conducting fillers: processing and properties. *Polym Compos,* **21**, 165–174.

35. Ten, E., Bahr, D. F., Li, B., Jiang, L., and Wolcott, M-P. (2012). Effects of cellulose nanowhiskers on mechanical, dielectric, and rheological properties of poly(3-hydroxybutyrate-co-3-hydroxyvalerate)/ cellulose nanowhisker composites. *Ind Eng Chem Res,* **51**, 2941–2951.

36. Tian, M., Yan, B., Yao, Y., Zhang, L., Nishi, T., and Ning, N. (2014). Largely improved actuation strain at low electric field of dielectric elastomer by combining disrupting hydrogen bonds with ionic conductivity. *J Mater Chem C,* **2**, 8388–8397.

37. Ning, N., Wang, Z., Yao, Y., Zhang, L., and Tian, M. (2015). Enhanced electromechanical performance of bio-based gelatin/glycerin dielectric elastomer by cellulose nanocrystals. *Carbohydr Polym,* **130**, 262–267.

38. Tobjork, D. and Osterbacka, R. (2011). Paper electronics. *Adv Mater,* **23**, 1935–1961.

39. Gaspar, D., Fernandes, S. N., de Oliveira, A. G., Fernandes, J. G., Grey, P., Pontes, R. V., Pereira, L., Martins, R., Godinho, M. H., and Fortunato, E. (2014). Nanocrystalline cellulose applied simultaneously as the gate dielectric and the substrate in flexible field effect transistors. *Nanotechnology,* **25**, 094008.

40. Ajayan, P., Stephan, O., Colliex, C., and Trauth, D. (1994). Aligned carbon nanotube arrays formed by cutting a polymer resin-nanotube composite. *Science,* **265**, 1212–1214.

41. Moniruzzaman, M. and Winey, K. I. (2006). Polymer nanocomposites containing carbon nanotubes. *Macromolecules,* **39**, 5194–5205.

42. Diomigi, C., Posati, T., Benfenati, V., Sagnella, A., Pistone, A., Bonetti, S., Ruani, G., Dinelli, F., Padeletti, G., Zamboni, R., and Muccini, M. (2014). A nanostructured conductive bio-composite of silk fibroin-single walled carbon nanotubes. *J Mater Chem B,* **2**, 1424–1431.

43. Wang, S. F., Sehn, L., Zhang, W. D., and Tong, Y. J. (2005). Preparation and mechanical properties of chitosan/carbon nanotubes composites. *Biomacromolecules,* **15**, 3067–3072.

44. Ke, G., Guan, W., Tang, C., Guan, W., Zeng, D., and Deng, F. (2007). Covalent functionalization of multiwalled carbon nanotubes with a low molecular weight chitosan. *Biomacromolecules,* **8**, 322–326.

45. Erdem, A., Muti, M., Karadeniz, H., Congur, G., and Canavar, E. (2012). Electrochemical monitoring of indicator-free DNA hybridization by carbon nanotubes-chitosan modified disposable graphite sensors. *Colloid Surf B Biointer,* **95**, 222–228.

46. Pauliukaite, R., Ghica, M. E., Fatibello-Filho, O., and Brett, C M. A. (2010). Electrochemical impedance studies of chitosan-modified electrodes for application in electrochemical sensors and biosensors. *Electrochim Acta,* **55**, 6239–6247.

47. Perez, C., Prokhorov, E., Luna-Bárcenas, G., González-Campos, J. B., Sanchez, I. C., Gonzalez-Hernandez, J., Duarte, M. E. M., and Villaseñor-Ortega, F. (2010). Relaxation phenomena in chitosan-carbon nanotube bionanocomposites. *J Nanostr Polym Nanocomp,* **6**, 61–67.

48. Hernández-Vargas, J., González-Campos, J. B., Lara-Romero, J., Prokhorov, E., Luna.Bárcenas, G., Aviña-Verduzco, J. A., and González-Hernández, J. C. (2014). Chitosan/MWCNTs-decorated with silver nanoparticle composites: dielectric and antibacterial characterization. *J Appl Polym Sci,* **131**, 40214.

49. Musto, P., Russo, P., Cimino, F., Acierno, D., Lupò, G., and Petrarca, C. (2015). Dielectric behavior of biopolymer based composites containing multi wall carbon nanotubes: effect of filler content and aspect ratio. *Eur Polym J,* **64**, 170–178.

50. Li, Q., Xue, Q., Hao, L., Gao, X., and Zhang, Q. (2008). Large dielectric constant of the chemically functionalized carbon nanotube/polymer composites, *Compos Sci Technol,* **68**, 2290–2296.

51. Ahmad, K., Pan, W., and Shi, S. L. (2006). Electrical conductivity and dielectric properties of multiwalled carbon nanotube and alumina composites, *Appl Phys Lett,* **89**, 133102-1-2-3.

52. Sahila, S. and Jayakumari, L. S. (2015). Development of in situ generated graphene/benzoxazine-epoxy nanocomposite for capacitor applications, *Polym Compos,* **36**, 1–7.

53. Dušan, G. and Ghillányová, K. (2010) Ceramic material classes, In *Ceramic oxides: Ceramic Science and Technology, Volume 2: Properties,* Eds. Riedel, R., and Chen, I. W., pp. 3–4, Wiley-VCH Verlag GmbH & Co. kGaA, Weinheim.

54. Prabunathan, P., Sethuraman, K., and Alagar, M. (2014). Development of bio-based F-SBA-l5 reinforced epoxy nanocomposites for low-k dielectric applications. *High Perform Polym,* **26**, 283–289.

55. Du, M., Guo, B., and Jia, J. (2010). Newly emerging applications of halloysite nanotubes: a review. *Polym Int,* **59**, 574–582.

56. Prashantha, K., Lacrampe, M. F., and Krawczak, P. (2013). Highly dispersed polyamide-11/halloysite nanocomposites: thermal, rheological, optical, dielectric, and mechanical properties. *J Appl Polym Sci,* **130**, 313–321.

57. Prashantha, K., Soulestin, J., Lacrampe, M. F., Krawczak, P., Dupin, G., Claes, M., and Tewari, A. (2010). Electrical and dielectric properties of multi-walled carbon nanotube filled polypropylene nanocomposites. *Polym Polym Compos,* **18**, 489–494.

58. Laghzizil, A., Elherch, N., Bouhaouss, A., Lorente, G., Coradin, T., Livage, J. (1983). Electrical behavior of hydroxyapatites $M_{10}(PO_4)6(OH)_2$ (M = Ca, Pb, Ba). *Mater Res Bull,* **36**, 953–962.

59. Karuppasamy, K., Linda, T., Thanikaikarasan, S., Balakumar, S., Mahalingam, T., Sebastian, P. J., and Shajan, X. S. (2013). Electrical and dielectric behavior of nano-bio ceramic filler incorporated polymer electrolytes for rechargeable lithium batteries. *J New Mat Electrochem Syst,* **16**, 115–120.

60. Vigneshwaran, N., Kumar, S., Kathe, A., Varadarajan, P., and Prasad, V. (2006). Functional finishing of cotton fabrics using zinc oxide-soluble starch nanocomposites. *Nanotechnology,* **17**, 5087.

61. Lin, W., Xu, Y., Huang, C. C., Ma, Y., Shannon, K. B., Chen, D. R., Huang, and Y. W. (2009). Toxicity of nano- and micro-sized ZnO particles in human lung epithelial cells. *J Nanopart Res,* **11**, 28–39.

62. Faupel, F., Zaporojtchenko, V., Strunskus, T., and Elbahri, M. (2010). Metal-polymer nanocomposites for functional applications. *Adv Eng Mater,* **12**, 1177–1190.

63. Hanemann, T. and Szabó, D. V. (2010). Polymer-nanoparticle composites: from synthesis to modern applications. *Materials,* **3**, 3468–3517.

64. Janotti, A. and Van de Walle, C. G. (2009). Fundamentals of zinc oxide as a semiconductor. *IOP Publ*, **72**, 126501.

65. Zhou, J., Xu, N. S., and Wang, Z. L. (2006). Dissolving behavior and stability of ZnO wires in biofluids: a study on biodegradability and biocompatibility of ZnO nanostructures. *Adv Mater*, **18**, 2432–2435.

66. Chen, X. and Mao, S. S. (2007). Titanium dioxide nanomaterials: synthesis, properties, modifications, and applications., *Chem Rev*, **107**, 2891–2959.

67. Rouhi, J., Mahmud, S., Naderi, N., Ooi, CH. R., and Mahmood, M. R. (2013). Physical properties of fish gelatin-based bio-nanocomposite films incorporated with ZnO nanorods. *Nanoscale Res Lett*, **8**, 364 (1–6).

68. Rahman, M. A. A., Mahmud, S., Alias, A. K., and Nor, A. F. M. (2013). Effect of nanorod zinc oxide on electrical and optical properties of starch-based polymer nanocomposites. *J Phys Sci*, **24**, 17–28.

69. Díez-Pascual, A. M. and Díez-Vicente, A. L. (2015). Nano-TiO_2 reinforced PEEK/PEI blends as biomaterials for load-bearing implant applications. *Appl Mater Interfaces*, **7**, 5561–5573.

70. Díez-Pascual, A. M., Naffakh, M., Gonzalez-Dominguez, J. M., Ansón, A., Martinez-Rubi, Y., Martinez, M. T., Simard, B., and Gómez, M. A. (2010). High performance PEEK/carbon nanotube composites compatibilized with polysulfones-II. mechanical and electrical properties. *Carbon*, **48**, 3500–3511.

71. Venkat, N., Ouchen, F., Singh, K. M., Smith, S. R., Joyce, D. M., Miller, T., Yaney, P. P., Grote, J. G., and Naik, R. R. (2011). Nanobiosystems: processing, characterization, and applications IV, In *Bio-Dielectrics Based on DNA-Ceramic Hybrid Films*, Eds. Kobayashi, N., Ouchen, F., and Rau, I., pp. 1–10, Proc. SPIE 8103, San Diego, California, USA.

72. Daniel, M. C. and Astruc, D. (2004). Gold nanoparticles: assembly, supramolecular chemistry, quantum-size-related properties, and applications toward biology, catalysis, and nanotechnology. *Chem Rev*, **104**, 293–346.

73. Averitt, R. D., Sarkar, D., and Halas, N. J. (1997). Plasmon resonance shifts of Au-Coated Au2S nanoshells: insight into multicomponent nanoparticle growth. *Phys Rev Lett*, **22**, 4217–4220.

74. Zhang, T., Wang, W., Zhang, D., Zhang, X., Ma, Y., Zhou, Y., and Qi, L. (2010). Biotemplated synthesis of gold nanoparticle-bacteria cellulose nanofiber nanocomposites and their application in biosensing. *Adv Funct Mater*, **20**, 1152–1160.

75. Gruber, S., Taylor, R. N. K., Scheel, H., Greil, P., and Zollfrank, C. (2011). Cellulose-biotemplated silica nanowires coated with a dense gold nanoparticle layer. *Mater Chem Phys,* **129**, 19–22.

76. Massaro, A., Spano, F., Cazzato, P., Cingolani, R., and Athanassiou, A. (2011). Innovative optical tactile sensor for robotic system by gold nanocomposite material. *Prog Electromagn Res,* **16**, 145–158.

77. Massaro, A., Spano, F., Cazzato, P., Cingolani, R., and Athanassiou, A. (2010). Real time optical pressure sensing for tactile detection using gold nanocomposite material. *Proceeding of MNE,* 121–122.

78. Rakic, A., Djurisic, A. B., Elazar, J. M., and Majewski, M. L. (1998). Optical properties of metallic films for vertical-cavity optoelectronic devices. *Appl Opt,* **37**, 5271–5283.

79. Chen, M. and Horn, R. G. (2007). Refractive index of sparse layers of absorbed gold nanoparticles. *J Coll Interf Sci,* **315**, 814–818.

80. Massaro, A., Pierantoni, L., and Rozzi, T. (2005). Development of the EM coupling in laminated multilayered 3D optical waveguides. *Intern J Num Model,* **18**, 237–253.

Chapter 8

Hybrid Polymer Nanocomposite Systems

Bin Li[a] and Wei-Hong Zhong[b]
[a]*Department of Mechanical Engineering, Wichita State University, 1845 Fairmount Street, Wichita, Kansas 67260-0133, USA*
[b]*School of Mechanical and Materials Engineering, Washington State University, 405 NE Spokane Street, Pullman, Washington 99164-2920, USA*
bin.li@wichita.edu, katie_zhong@wsu.edu

Dispersion, distribution, and surface properties of different nanomaterials are among the most important factors that affect properties and functionalities of polymer nanocomposites. Therefore, a great amount of effort has been made to realize high-level uniform dispersion and precisely controlled distribution and surface properties of nanomaterials, in order to obtain desired dielectric properties, as discussed in the previous chapters.

In addition to these efforts on a traditional composites, there are many new nanocomposite structures providing much potential for enhanced dielectric performances, thanks to the fast development of materials chemistry and physics, nanotechnology, and polymer processing techniques [2, 3, 5–10, 12, 14–17, 20, 22, 24–32, 34–40, 42–48, 51–69]. The most popular and successful new structures

Polymer Nanocomposites for Dielectrics
Edited by Wei-Hong Zhong and Bin Li

ISBN 978-981-4745-53-6 (Hardcover), 978-981-4745-54-3 (eBook)
www.panstanford.com

are polymer nanocomposites with hybrid nanomaterial systems. Hybrid polymer nanocomposites not only have improved dielectric permittivity and energy storage capability for the polymer matrices, but have also shown incredible effectiveness in preventing large increase in energy loss. These unconventional nanocomposites have more complex structures and compositions. In addition, this complexity also provides more controllable structural and material parameters for the customization of dielectric properties.

8.1 Hybrid Systems with Isolated Nanomaterials

Hybrid nanomaterials systems consist of more than one type of nanomaterials or nanostructures. A certain synergy between different nanomaterials in the same system is typically expected, in order to achieve properties better than those of the traditional nanocomposites [23]. By directly mixing a polymer matrix with different types of nanomaterials, a hybrid polymer nanocomposite structure can be simply fabricated. A common synergistic effect is the improvement of dispersion quality of nanomaterials, leading to enhancement in the properties and functionalities. However, the understanding of this effect is very limited today.

Figure 8.1 is an example of polymer nanocomposites made by directly mixing $BaTiO_3$ nanoparticles (NBTs) and multiwalled carbon nanotubes (MWCNTs) with PVDF [6]. It is clear that the phase structures of this ternary polymer nanocomposite were related to the compositions. The resulting dielectric properties showed a visible dependence on the concentrations of both NBTs and MWCNTs. In particular, at a volume fraction of 0.02 MWCNT, a high dielectric constant was achieved, regardless of the concentration of NBTs. At the same time, it was found that the dielectric loss was mostly from conduction loss. This composition-dependent dielectric constant could be understood by the evolution of the network structures of MWCNTs and NBTs in PVDF matrix with the variation of the compositions. According to the TEM images in Fig. 8.1, the dispersion of MWCNTs was improved with the addition of NBTs. In particular, the high volume fraction of NBTs seemed to benefit the dispersion of MWCNTs by individualizing bundled MWCNTs and favor the formation of the MWCNT network. The uniform dispersion leads to a large number of microcapacitor structures shown in the dashed circles in Fig. 8.1. Similar isolating effects of NBTs were also found in the PVDF/graphene/NBT nanocomposites [41].

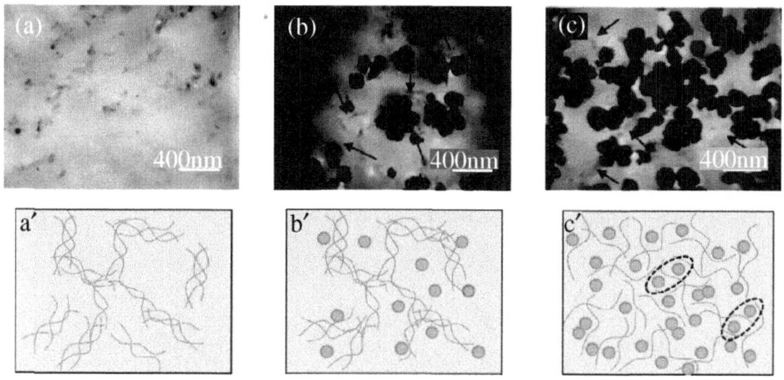

Figure 8.1 TEM images of the freeze-fractured (NBT-MWCNT)/PVDF nanocomposites with a constant f_{MWCNT} of 0.02 at (a) $f_{NBT} = 0$, (b) $f_{NBT} = 0.05$, (c) $f_{NBT} = 0.20$, and the schematic illustration of the microstructure of the (NBT-MWCNT)/PVDF composites (a′) $f_{NBT} = 0$, (b′) $f_{NBT} = 0.05$, (c′) $f_{NBT} = 0.20$, in which the gray lines stand for the MWCNTs and the green balls are NBTs. Reprinted with permission from Ref. [6], Copyright 2010, American Chemical Society.

Although the mechanisms of this improved dispersion are still unclear, the enhancement in the dielectric constant suggests an alternative approach to traditional nanocomposites. Meanwhile, when the volume fraction of NBTs is above 0.05, the reduction in dielectric loss does not lead to a decrease in the dielectric constant, showing the potential of the hybrid polymer nanocomposites for dielectric applications requiring high dielectric constant and low dielectric loss. As a matter of fact, the hybrid polymer nanocomposites often showed reduced dielectric loss, while the dielectric constant maintained at high levels.

Boron nitride has high breakdown strength, which is superior to that of BTs. In the study shown in Fig. 8.3 [25], the boron nitride nanosheets (BNNS) were added to P(VDF-CTFE)/BT to form nanocomposites, in order to improve the energy storage performances. The effect of BNNS on the dispersion of BT nanoparticles was observed. With the addition of BNNS, the dispersion quality of BTs was improved, which was postulated to be related to the 2D nanostructures of BNNSs with high specific surface area. The resulting ternary polymer nanocomposites with improved dispersion of BTs by addition of BNNS possessed large discharged energy density, high charge–discharge efficiency, and great power density. Other isolated hybrid nanomaterial systems incorporated into polymer matrix include BT/SiC [65], BT/MWCNT [15, 55], Al/β-SiC_W [66], and so on. Both high dielectric constant and low dielectric loss were observed in these nanocomposites.

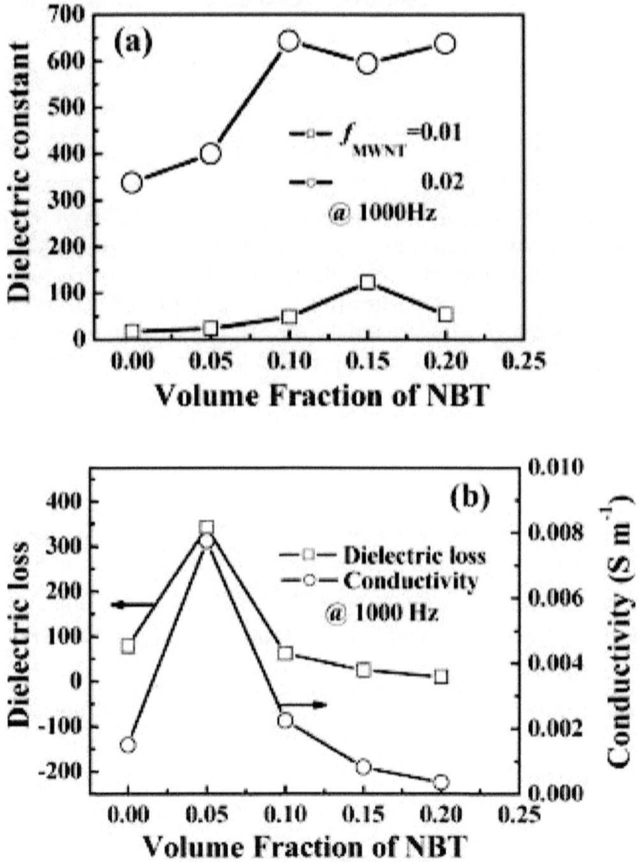

Figure 8.2 Concentration dependent dielectric and electrical properties of (NBT-MWCNT)/PVDF composites with an f_{MWCNT} of 0.02 at 1000 Hz. Reprinted with permission from Ref. [6], Copyright 2010, American Chemical Society.

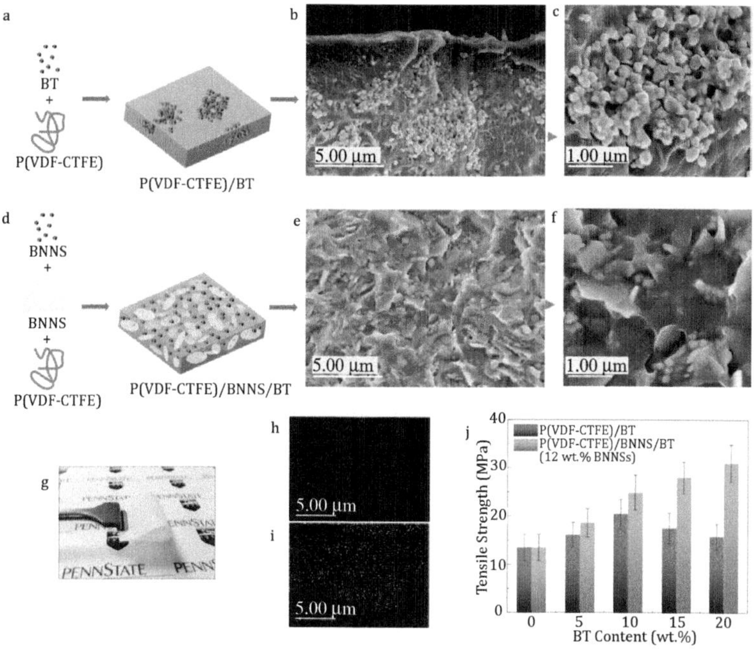

Figure 8.3 Schematic and cross-sectional SEM images of P(VDF-CTFE)/BT composite film with 15 wt% BT, and of P(VDF-CTFE)/BNNS/BT ternary nanocomposite film with 12 wt% BNNSs and 15 wt% of BT. Reprinted from Ref. [25], with permission from John Wiley and Sons.

8.2 Covalently Bonded Hybrid Nanomaterials

The attractiveness of the hybrid nanomaterials is that, in addition to directly blending of different species of nanomaterials, various hybrid nanostructures can be realized via different chemical approaches, such as the covalently bonded nanostructures as shown in Figs. 8.4 [40] and 8.5 [37]. In the study of the hybrid nanocomposites of PVDF and BT and graphene oxide (GO), the nanoparticles of BT and graphene were covalently bonded via the reaction between γ-aminopropyltriethoxysilane (APTS)-modified BT and GO. Compared with PVDF/GO and PVDF/BT nanocomposites, the hybrid nanocomposite demonstrated very high dielectric constant, suggesting the potential for achieving high energy storage capability.

In another study shown in Fig. 8.5, BT nanoparticles were covalently linked to organomodified montmorillonite (oMMT) nanoparticles. According to Fig. 8.5, without the covalent links between oMMT and BT nanoparticles, the oMMT dispersed in epoxy mainly in the form of aggregates, while the BT nanoparticles showed good dispersion. However, the 2D oMMT has much bigger surface area, attributing to the interface between the polymer matrix and nanoparticles. Therefore, the poor dispersion of oMMT in epoxy did not contribute to the properties of the nanocomposites in a positive manner. This issue was resolved after the BT covalently reacted with oMMT. Obviously, the dispersion of oMMT was significantly improved with the uniformly dispersed oMMTs surrounding the BT nanoparticles. This nanocomposite structure enhanced the interfaces in the polymer nanocomposites. Meanwhile, due to the high breakdown strength of oMMT, the uniformly dispersed oMMT also improved the breakdown strength of the resulting nanocomposites.

According to Fig. 8.6 [37], the hybrid nanocomposites possessed remarkably enhanced performance compared to the pure epoxy (curve a), as well as better performance than that of the composites with similar concentration of nonreactive oMMT (curve b). In particular, the enhanced dielectric polarization did not lead to a higher energy loss, which is desirable for energy storage applications.

Both isolated and covalently bonded hybrid nanomaterials contribute to better dispersion quality of nanomaterials in the polymer matrix. The advantages of both approaches are obvious. First of all, the hybrid polymer nanocomposites can benefit from excellent and unique properties of each component in the hybrid nanomaterials, such as the high breakdown strength of oMMT and high permittivity of BT in Figs. 8.5 and 8.6. Secondly, the improved dispersion generates large interfaces between polymer and nanomaterials, benefiting the properties and functionalities related to interfacial structures and properties, such as mechanical properties and dielectric properties. Also, for isolated hybrid nanomaterials, the improved dispersion was realized without extra surface modification processes, which is positive to the graphitic nanomaterials, and the surface modification often causes structural damages.

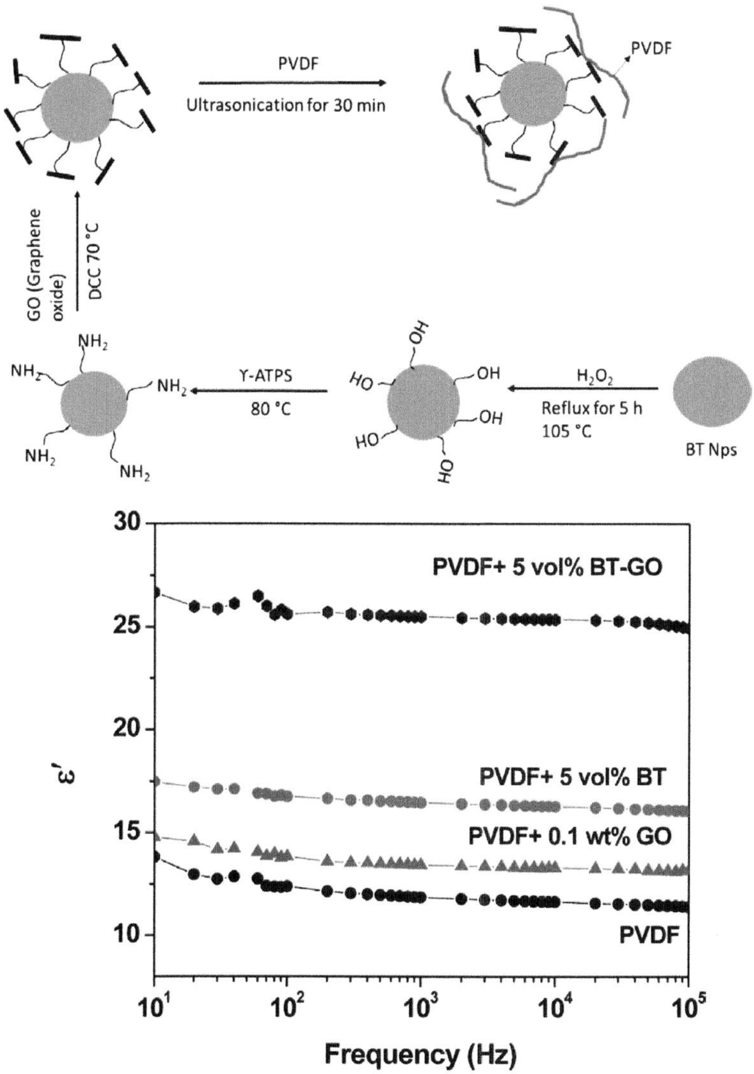

Figure 8.4 Preparation of PVDF/CNT/BT–GO composites and their dielectric constant. Reprinted with permission from Ref. [40], Copyright 2014, American Chemical Society.

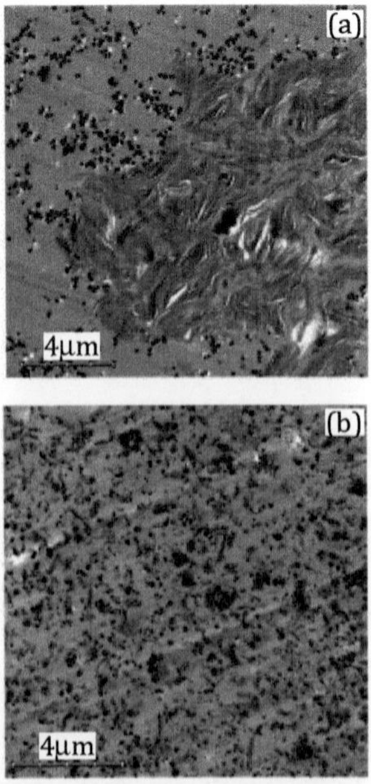

Figure 8.5 TEM images of the $BaTiO_3$/oMMT-based nanocomposites: (a) without covalent bond between BT and oMMT, at 10 vol% BT and 3 wt% oMMT; (b) with 20 wt% hybrid fillers, i.e., BT covalently bonded with reactive oMMT at 2:1 BT:oMMT ratio. The two composites shown in (a) and (b) have essentially the same components and differ only in the filler bonding. Reprinted with permission from Ref. [37], Copyright 2010, AIP Publishing LLC.

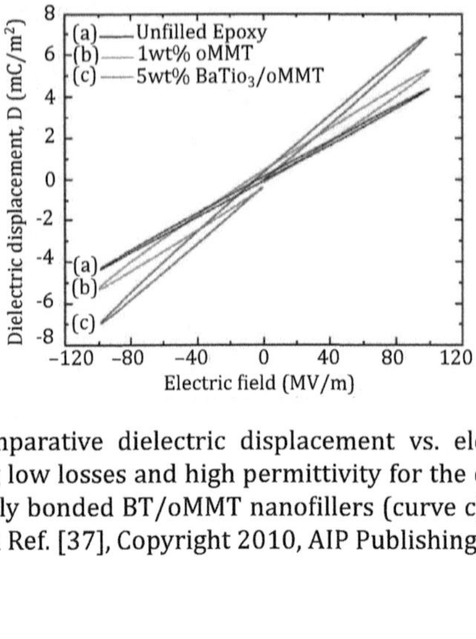

Figure 8.6 Comparative dielectric displacement vs. electric field (*D–E* plots) indicating low losses and high permittivity for the composite with 5 wt% in covalently bonded BT/oMMT nanofillers (curve c). Reprinted with permission from Ref. [37], Copyright 2010, AIP Publishing LLC.

8.3 Core–Shell Hybrid Nanostructures

Core–shell structures are one of the most important material structures studied in the field of composite materials [1, 11, 18, 19, 33, 50]. In general, we can consider all surface-modified nanomaterials as core–shell nanostructures with a surfactant shell covering the surface of a nanomaterial core. In Chapter 3, we discussed on the core–shell structures with polymer shell-inorganic cores [48, 49], in which the structures and properties have shown significant impacts on the dielectric properties of the resulting nanocomposites. Core–shell structures could be created by combining any two types of materials, for example, ceramic shell–ceramic core [21], ceramic shell–graphitic carbon core [13, 16], metal core–ceramic shell [4], and so on. In addition to the properties of both shell and core materials, the shell thickness and the size of the core are also critical to the performances of these core–shell nanostructures.

Figure 8.7 presents the contribution of ZnO@Zn core–shell nanostructures to the dielectric properties of PVDF nanocomposites [63]. With the increasing volume fraction of ZnO@Zn nanoparticles, the dielectric constant increased, while the dielectric loss only exhibited small increase. The relative thickness of the ZnO shell also showed significant effects on the dielectric properties. The thickness of ZnO shell was controlled by calcination time and temperature. In this study, the high calcination temperature and long calcination time led to the thicker ZnO layer. Dielectric constant increased with the thickness of the ZnO layer. At the same time, the highest dielectric constant was found in PVDF/ZnO@Zn nanocomposites. ZnO is a typical dielectric ceramic showing high polarization. The increasing content of ZnO contributes to the improvement in dielectric performance, according to the effective medium theories. However, the highest dielectric constant occurred to the hybrid nanocomposites. A possible reason for the dielectric properties was considered a result of duplex polarizations caused by two interfaces in this nanocomposite. At the same time, with the obvious increase in dielectric constant, the dielectric loss only slightly went up. This should be associated with the changes of interfacial properties after the ZnO layer was introduced. The dielectric loss of composites is related to electrical properties as well as the interfacial polarization mechanisms. The ZnO is a semiconductor, which could substantially change the interface structure and properties between PVDF matrix and Zn nanoparticles.

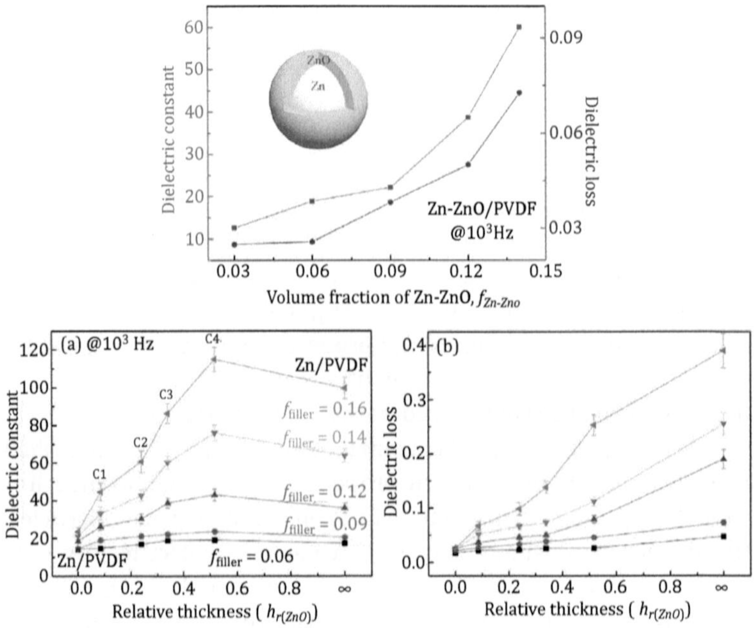

Figure 8.7 Volume fraction–dependent dielectric properties of PVDF/ZnO@Zn nanocomposites and the effects of relative thickness of ZnO coating. Reprinted with permission from Ref. [63], Copyright 2012, American Chemical Society.

From this representative research on the core–shell nanostructure, we can clearly see that dielectric properties of this type of hybrid polymer nanocomposites strongly depend on the core–shell structures of the nanomaterials, i.e., the relative thickness of shell layer and core structure. This is very critical to the interfacial structures and polarization processes, impacting the dielectric properties of the hybrid nanocomposite at large. For example, contrary to the research shown in Fig. 8.7, the TiO_2/Ag-modified PVDF nanocomposites exhibited clear shell thickness–dependent dielectric properties; on the other hand, the thicker TiO_2 shell favored a lower dielectric constant [7]. Today, the understanding on this complex interfacial polarization in the polymer nanocomposites modified by core–shell nanostructures is still in its infant state. A few factors should be investigated in order to understand this complex interfacial phenomena, such as the dielectric and electric properties of shell and core materials and the synergy between the shell and core structures.

Low dielectric loss seems to be a characteristic property of many hybrid nanocomposites of this kind, in particular the nanocomposites with a core–shell structure consisting of a conductive core and an insulating/semiconducting shell [53]. As discussed in Chapter 4, the interfacial polarization between conductive graphitic nanomaterials and polymer matrix usually produced very high dielectric constant, which was accompanied with amazingly high dielectric loss. One major reason for this very high dielectric loss was because of the high electrical conduction. By introducing an insulating or semiconducting layer on the surface of these conductive nanomaterials, the dielectric loss could be restrained, probably due to the reduced conduction loss as a result of weakened electron tunneling at the interfaces [9, 45, 53].

This core–shell structure can also be developed to nanofibers, as shown in Fig. 8.8 [28]. The existence of the SiO_2 nanolayer on the surface of BT nanofibers led to the increased discharged energy density and high energy efficiency. As discussed in Chapter 2, for energy storage applications, the discharged energy is very important. In many polymer nanocomposites with high energy storage density, due to very large energy loss, the discharged energy is very limited. SiO_2 in this nanostructure is a good electrical insulator, which could block some charge carrier movements in the resulting nanocomposites. This will lead to weaker interfacial polarization with PVDF; however, at the same time, it favors the reduction of dielectric loss, contributing to very high discharged energy density. In other words, the resulting PVDF nanocomposites can have high energy efficiency applications.

Semiconductors and insulators can significantly alter the electrical and dielectric properties at the polymer–nanomaterial interfaces. Particularly, the movement of charge carriers near the interfacial areas will be dramatically affected by these semiconducting and insulating layers. This effect will benefit low dielectric loss, but at the same time, the slight sacrifice of high dielectric constant or energy storage capability is often needed, due to the polarization mechanisms. Therefore, how to optimize the core–shell structures to achieve both high energy storage capability and very low dielectric loss is one major challenge to the applications of this class of hybrid nanocomposites.

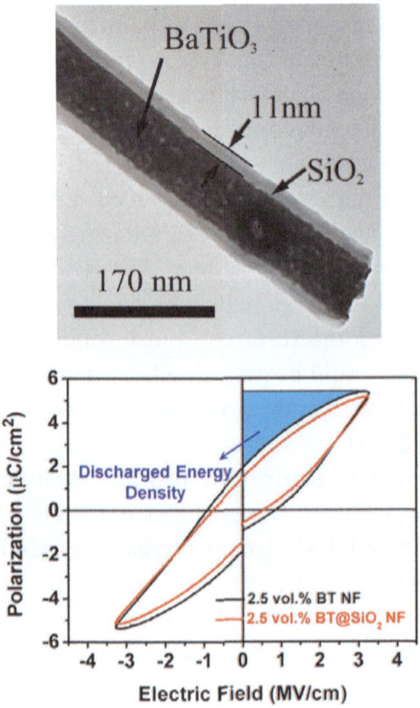

Figure 8.8 TEM morphology of the core-shell structured BT@SiO_2 nanofibers, and discharged energy density and energy storage efficiency of composites of the nanocomposite of BT@SiO_2 NF and that of BT NF with a concentration of 2.5 vol%, as functions of the applied electric field. Reprinted with permission from Ref. [28], Copyright 2015, AIP Publishing LLC.

The hybrid nanostructures in Fig. 8.9 can be treated as a core–shell structure with only partial shell structures [31]. The Ag nanoparticles were directly synthesized on the surface of BT nanoparticles. This process could prevent aggregation of Ag nanoparticles; therefore, the enhanced interfacial polarization between polymer materials and Ag nanoparticles could be achieved. During the preparation of nanocomposites, both mechanical stirring and ultrasonication were applied. However, the Ag nanoparticles were still attached to the surface of BT nanoparticles, suggesting a very good adhesion. With the increase of concentration of the BT–Ag nanomaterials, both dielectric constant and electrical conductivity increased. The interesting aspect of this nanostructure is that because of the separately deposited Ag nanoparticles, there are three types of interfaces: polymer–Ag, polymer–BT, and BT–Ag, contributing to a more complex interfacial polarization phenomena. Knowledge of this subject will benefit the applications of this type of hybrid nanostructures in advanced nanocomposites.

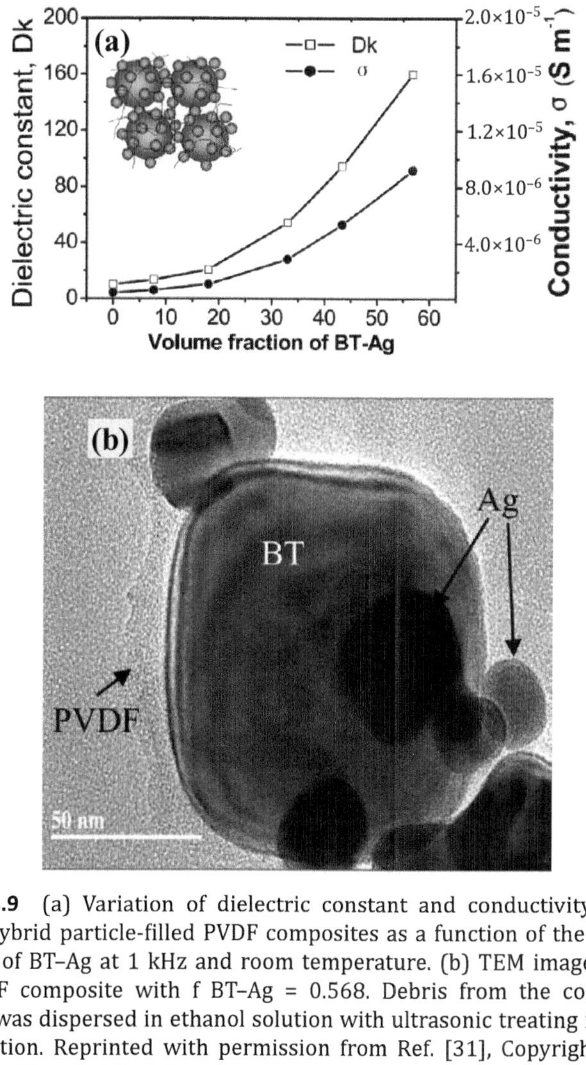

Figure 8.9 (a) Variation of dielectric constant and conductivity of the BT–Ag hybrid particle-filled PVDF composites as a function of the volume fraction of BT–Ag at 1 kHz and room temperature. (b) TEM image of BT–Ag/PVDF composite with f BT–Ag = 0.568. Debris from the composite sample was dispersed in ethanol solution with ultrasonic treating for TEM examination. Reprinted with permission from Ref. [31], Copyright 2014, American Chemical Society.

8.4 Composite Nanomaterials

A hybrid nanomaterial system can be a composite itself, such as the nanofiber structures in Fig. 8.10 [60]. In this structure, the TiO_2 nanoparticles were dispersed in $BaTiO_3$ matrix, forming ceramic–ceramic composite nanofibers. The composite nanofibers were incorporated into the PVDF matrix. The resulting PVDF/ BTO@ TO_nfs nanocomposites showed discharged energy density much superior to PVDF/BTO_nfs nanocomposites and PVDF/TO_nfs nanocomposites. In particular, with only 3 vol% of nanofibers in PVDF, the discharged energy density was as high as 20 J/cm^3. This remarkable enhancement was believed to have resulted from the large interfacial area between $BaTiO_3$ and TiO_2 nanoparticles. In a similar system, the core–shell structure ($BaTiO_3$@TiO_2) consisting of TiO_2 shell and $BaTiO_3$ core was incorporated into P(VDF-HFP). Although the enhancement in the discharged energy density is obvious compared with pure polymer matrix (12.2 J/cm^3), first of all, it is still interior to the BTO@TO nanofibers; secondly, a large amount of core–shell nanoparticles was needed [38].

This comparison revealed the significance of hybrid nanostructures to the dielectric performances of polymer nanocomposites, since it accounted for different interfacial polarization mechanisms. Among many factors affecting the different energy storage capabilities, it is believed that the composite nanofiber structures have stronger interfacial polarization between TiO_2 and $BaTiO_3$. Meanwhile, the 1D structures typically favor larger specific surface area compared with 0D structures. Therefore, the stronger interfacial polarization between a polymer matrix and nanomaterials will be expected. Combined with all the hybrid nanostructures discussed in this chapter, it is obvious that there are many controllable structural parameters in hybrid nanostructures compared with conventional nanostructures. Considering their great potential for energy applications requiring high energy capability and low energy loss, systematic and comprehensive studies on these parameters are in high demand. An in-depth understanding of the relationships between dielectric properties and these controllable parameters will also advance our knowledge in dielectric polarization theories in multiphase material structures.

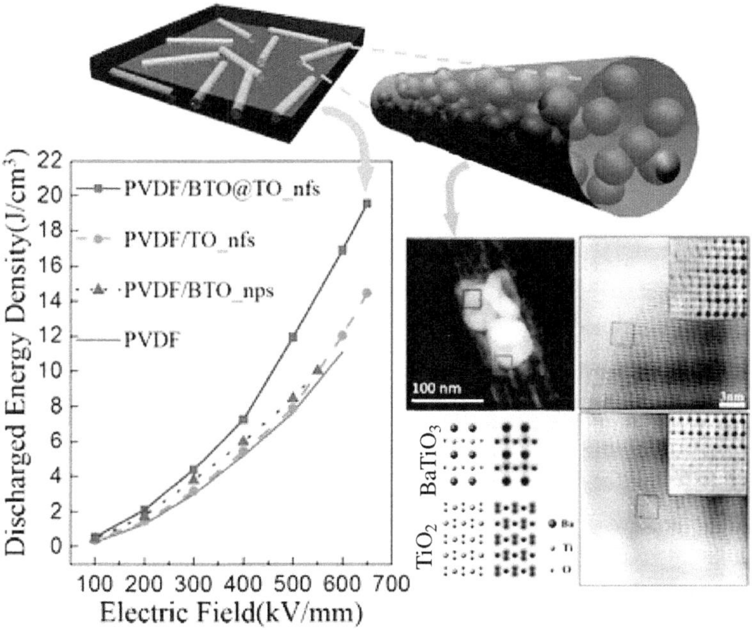

Figure 8.10 Schematic illustration of BTO@TO_nfs and PVDF/BTO@TO_nfs nanocomposites. Discharged energy density of PVDF nanocomposites embedded with BTO@TO_nfs, TO_nfs, BTO_nps, and pure PVDF films as a function of electric field. The volume fraction of the three nano-inclusions were fixed at 3% in all composites. Reprinted from Ref. [60], with permission from John Wiley and Sons.

References

1. Buscaglia, M. T., Sennour, M., Buscaglia, V., Bottino, C., Kalyani, V., and Nanni, P. (2011). Formation of $Bi_4Ti_3O_{12}$ one-dimensional structures by solid-state reactive diffusion: from core-shell templates to nanorods and nanotubes. *Cryst Growth Des*, **11**, 1394–1401.

2. Carr, J. M., Mackey, M., Flandin, L., Schuele, D., Zhu, L., and Baer, E. (2013). Effect of biaxial orientation on dielectric and breakdown properties of poly(ethylene terephthalate)/poly(vinylidene fluoride-co-tetrafluoroethylene) multilayer films. *J Polym Sci, Part B:Polym Phys*, **51**, 882–896.

3. Chen, G. X., Zhang, S. C., Zhou, Z., and Li, Q. F. (2015). Dielectric properties of poly(vinylidene fluoride) composites based on bucky gels of carbon nanotubes with ionic liquids. *Polym Compos*, **36**, 94–101.

4. Choi, S. H., Ko, Y. N., Lee, J. K., Park, B. K., and Kang, Y. C. (2013). Core-shell-structure Ag-SaTiO(3) composite nanopowders prepared directly by flame spray pyrolysis. *Mater Chem Phys*, **140**, 266–272.
5. Chu, L. Y., Xue, Q. Z., Sun, J., Xia, F. J., Xing, W., Xia, D., and Dong, M. D. (2013). Porous graphene sandwich/poly(vinylidene fluoride) composites with high dielectric properties. *Compos Sci Technol*, **86**, 70–75.
6. Dang, Z. M., Yao, S. H., Yuan, J. K., and Bai, J. B. (2010). Tailored dielectric properties based on microstructure change in $BaTiO_3$-carbon nanotube/poly(vinylidene fluoride) three-phase nanocomposites. *J Phys Chem C*, **114**, 13204–13209.
7. Dang, Z. M., You, S. S., Zha, J. W., Song, H. T., and Li, S. T. (2010). Effect of shell-layer thickness on dielectric properties in Ag@TiO_2 core@shell nanoparticles filled ferroelectric poly(vinylidene fluoride) composites. *Phys Status Solidi A*, **207**, 739–742.
8. Fan, P., Wang, L., Yang, J. T., Chen, F., and Zhong, M. Q. (2012). Graphene/poly(vinylidene fluoride) composites with high dielectric constant and low percolation threshold. *Nanotechnology*, **23**.
9. Feng, M. N., Huang, X., Pu, Z. J., and Liu, X. B. (2014). Dielectric and mechanical properties of three-component Al_2O_3/MWCNTs/polyarylene ether nitrile micro-nanocomposite. *J Mater Sci: Mater Electron*, **25**, 1393–1399.
10. Goyal, R. K. and Kulkarni, A. B. (2012). Electrical properties of novel three-phase polymer nanocomposites with a high dielectric constant. *J Phys D: Appl Phys*, **45**.
11. Gupta, V., Patra, M. K., Shukla, A., Saini, L., Songara, S., Jani, R., Vadera, S. R., and Kumar, N. (2014). Synthesis and investigations on microwave absorption properties of core-shell FeCo(C) alloy nanoparticles. *Sci Adv Mater*, **6**, 1196–1202.
12. Hu, P. H., Shen, Y., Guan, Y. H., Zhang, X. H., Lin, Y. H., Zhang, Q. M., and Nan, C. W. (2014). Topological-structure modulated polymer nanocomposites exhibiting highly enhanced dielectric strength and energy density. *Adv Funct Mater*, **24**, 3172–3178.
13. Huang, L., Zhu, P. L., Li, G., Lu, D. Q., Sun, R., and Wong, C. P. (2014). Core-shell SiO_2@RGO hybrids for epoxy composites with low percolation threshold and enhanced thermo-mechanical properties. *J Mater Chem A*, **2**, 18246–18255.
14. Huang, X., Pu, Z. J., Feng, M. N., Tong, L. F., and Liu, X. B. (2013). $BaTiO_3$@MWCNTs core/shell nanotubes embedded PEN nanocomposite films

with high thermal stability and high permittivity. *Mater Lett*, **96**, 139–142.

15. Huang, X., Pu, Z. J., Tong, L. F., Zhao, R., and Liu, X. B. (2013). Novel PEN/$BaTiO^3$/MWCNT multicomponent nanocomposite film with high thermal stability for capacitor applications. *J Electron Mater*, **42**, 726–733.
16. Huang, X., Feng, M. N., and Liu, X. B. (2014). Synergistic enhancement of dielectric constant of novel core/shell $BaTiO^3$@MWCNTs/PEN nanocomposites with high thermal stability. *J Mater Sci:Mater Electron*, **25**, 97–102.
17. Jilani, S. M. and Banerji, P. (2014). Effect of ZnO loading on the electrical characteristics of graphene oxide-ZnO based thin film transistors. *Phys Semicond Dev*, 615–616.
18. Klonos, P., Kyritsis, A., and Pissis, P. (2015). Effects of surface modification and thermal annealing on the interfacial dynamics in core-shell nanocomposites based on silica and adsorbed PDMS. *Eur Polym J*, **70**, 342–359.
19. Klonos, P., Sulym, I. Y., Kyriakos, K., Vangelidis, I., Zidropoulos, S., Sternik, D., Borysenko, M. V., Kyritsis, A., Derylo-Marczewska, A., Gun'ko, V. M., and Pissis, P. (2015). Interfacial phenomena in core-shell nanocomposites of PDMS adsorbed onto low specific surface area fumed silica nanooxides: Effects of surface modification. *Polymer*, **68**, 158–167.
20. Kuang, X. W., Liu, Z., and Zhu, H. (2013). Dielectric properties of Ag@C/PVDF composites. *J Appl Polym Sci*, **129**, 3411–3416.
21. Laguta, V. V., Elissalde, C., Maglione, M., Artemenko, A. M., Chlan, V., Stepankova, H., and Zagorodniy, Y. (2015). Crystal structure transformations induced by surface stresses in $BaTiO_3$ and $BaTiO_3$@SiO_2 nanoparticles and ceramics. *Phase Transitions*, **88**, 761–775.
22. Lean, M. H., Wolak, M. A., Mackey, M., and Baer, E. (2014). Internal field distributions in multilayer polycarbonate/poly(vinylidene fluoride)-hexafluoropropylene films at onset of breakdown. *IEEE Trans Dielectr Electr Insul*, **21**, 800–808.
23. Li, B. and Zhong, W. H. (2011). Review on polymer/graphite nanoplatelet nanocomposites. *J Mater Sci*, **46**, 5595–5614.
24. Li, G., Yu, S. H., Sun, R., and Lu, D. (2015). Clean and in-situ synthesis of copper-epoxy nanocomposite as a matrix for dielectric composites with improved dielectric performance. *Compos Sci Technol*, **110**, 95–102.

25. Li, Q., Han, K., Gadinski, M. R., Zhang, G. Z., and Wang, Q. (2014). High energy and power density capacitors from solution-processed ternary ferroelectric polymer nanocomposites. *Adv Mater*, **26**, 6244–6249.

26. Li, Y., Huang, X., Hu, Z., Jiang, P., Li, S., and Tanaka, T. (2011). Large dielectric constant and high thermal conductivity in poly(vinylidene fluoride)/barium titanate/silicon carbide three-phase nanocomposites. *ACS Appl Mater Interfaces*, **3**, 4396–4403.

27. Lin, Y., Shaffer, J. W., and Sodano, H. A. (2010). Electrolytic deposition of PZT on carbon fibers for fabricating multifunctional composites. *Smart Mater Struct*, **19**.

28. Liu, S. H., Xue, S. X., Shen, B., and Zhai, J. W. (2015). Reduced energy loss in poly(vinylidene fluoride) nanocomposites by filling with a small loading of core- shell structured $BaTiO_3/SiO_2$ nanofibers. *Appl Phys Lett*, **107**.

29. Lockhart, R., Dauksevicius, R., Quintero, A. V., Janphuang, P., Briand, D., and de Rooij, N. F. (2013). Flexible and robust multilayer micro-vibrational harvesters for high acceleration environments. *13th International Conference on Micro and Nanotechnology for Power Generation and Energy Conversion Applications (Powermems 2013)*, **476**.

30. Long, Y. and Liu, X. B. (2014). Dielectric, mechanical and thermal properties of novel core-shell CuPc@MWCNTs/PEN composite films. *J Mater Sci: Mater Electron*, **25**, 1089–1096.

31. Luo, S. B., Yu, S. H., Sun, R., and Wong, C. P. (2014). Nano Ag-deposited $BaTiO_3$ hybrid particles as fillers for polymeric dielectric composites: toward high dielectric constant and suppressed loss. *ACS Appl Mater Interfaces*, **6**, 176–182.

32. Mackey, M., Schuele, D. E., Zhu, L., Flandin, L., Wolak, M. A., Shirk, J. S., Hiltner, A., and Baer, E. (2012). Reduction of dielectric hysteresis in multi layered films via nanoconfinement. *Macromolecules*, **45**, 1954–1962.

33. Ohlan, A., Singh, K., Chandra, A., and Dhawan, S. K. (2010). Microwave absorption behavior of core-shell structured poly (3,4-ethylenedioxy thiophene)-barium ferrite nanocomposites. *ACS Appl Mater Interfaces*, **2**, 927–933.

34. Ounaies, Z., Park, C., Harrison, J., and Lillehei, P. (2008). Evidence of piezoelectricity in SWNT-polyimide and SWNT-PZT-polyimide composites. *J Thermoplast Compos Mater*, **21**, 393–409.

35. Pal, K., Kang, D. J., Zhang, Z. X., and Kim, J. K. (2010). Synergistic effects of zirconia-coated carbon nanotube on crystalline structure of

poly(vinylidene fluoride) nanocomposites: electrical properties and flame-retardant behavior. *Langmuir*, **26**, 3609–3614.

36. Paleo, A. J., Martinez-Boubeta, C., Balcells, L., Costa, C. M., Sencadas, V., and Lanceros-Mendez, S. (2011). Thermal, dielectrical and mechanical response of alpha and beta-poly(vinilydene fluoride)/Co-MgO nanocomposites. *Nanoscale Res Lett*, **6**.

37. Polizos, G., Tomer, V., Manias, E., and Randall, C. A. (2010). Epoxy-based nanocomposites for electrical energy storage. II: Nanocomposites with nanofillers of reactive montmorillonite covalently-bonded with barium titanate. *J Appl Phys*, **108**, 074117.

38. Rahimabady, M., Mirshekarloo, M. S., Yao, K., and Lu, L. (2013). Dielectric behaviors and high energy storage density of nanocomposites with core-shell $BaTiO_3$@TiO_2 in poly(vinylidene fluoride-hexafluoropropylene). *Phys Chem Chem Phys*, **15**, 16242–16248.

39. Ren, L., Zhao, J., Wang, S. J., Zha, J. W., Hu, G. H., and Dang, Z. M. (2015). Remarkably variable dielectric and magnetic properties of poly(vinylidene fluoride) nanocomposite films with triple-layer structure. *Compos Sci Technol*, **107**, 107–112.

40. Sharma, M., Singh, M. P., Srivastava, C., Madras, G., and Bose, S. (2014). Poly(vinylidene fluoride)-based flexible and lightweight materials for attenuating microwave radiations. *ACS Appl Mater Interfaces*, **6**, 21151–21160.

41. Shen, Y., Guan, Y. H., Hu, Y. H., Lei, Y. C., Song, Y., Lin, Y. H., and Nan, C. W. (2013). Dielectric behavior of graphene/$BaTiO_3$/poly(vinylidene fluoride) nanocomposite under high electric field. *Appl Phys Lett*, **103**.

42. Singh, S., Kumar, N., Jha, A., Sahni, M., Sung, K. D., Jung, J. H., and Chaubey, S. (2015). Study of magnetic, dielectric and magnetodielectric properties of $BaTiO_3$/Fe_3O_4 core/shell nanocomposite. *J Mater Sci: Mater Electron*, **26**, 32–36.

43. Tseng, J. K., Tang, S., Zhou, Z., Mackey, M., Carr, J. M., Mu, R., Flandin, L., Schuele, D. E., Baer, E., and Zhu, L. (2014). Interfacial polarization and layer thickness effect on electrical insulation in multilayered polysulfone/poly(vinylidene fluoride) films. *Polymer*, **55**, 8–14.

44. Wang, S. F., Wang, Y. R., Cheng, K. C., and Chen, S. H. (2010). Physical and electrical properties of polyimide/ceramic hybrid films prepared via non-hydrolytic sol-gel process. *J Mater Sci: Mater Electron*, **21**, 104–110.

45. Wang, Z. J., Zhou, W. Y., Sui, X. Z., and Dong, L. N. (2015). Enhanced dielectric properties and thermal conductivity of Al/CNTs/PVDF ternary composites. *J Reinf Plast Compos*, **34**, 1126–1135.

46. Wolak, M. A., Wan, A. S., Shirk, J. S., Mackey, M., Hiltner, A., and Baer, E. (2012). Imaging the effect of dielectric breakdown in a multilayered polymer film. *J Appl Polym Sci*, **123**, 2548–2557.

47. Wu, L. W., Wang, X. H., Gong, H. L., Hao, Y. N., Shen, Z. B., and Li, L. T. (2015). Core-satellite $BaTiO_3$@$SrTiO_3$ assemblies for a local compositionally graded relaxor ferroelectric capacitor with enhanced energy storage density and high energy efficiency. *J Mater Chem C*, **3**, 750–758.

48. Xie, L., Huang, X., Huang, Y., Yang, K., and Jiang, P. (2013). Core@double-shell structured $BaTiO_3$–polymer nanocomposites with high dielectric constant and low dielectric loss for energy storage application. *J Phys Chem C*, **117**, 22525–22537.

49. Xie, L. Y., Huang, X. Y., Huang, Y. H., Yang, K., and Jiang, P. K. (2013). Core-shell structured hyperbranched aromatic polyamide/ $BaTiO_3$ hybrid filler for poly(vinylidene fluoride-trifluoroethylene-chlorofluoroethylene) nanocomposites with the dielectric constant comparable to that of percolative composites. *ACS Appl Mater Interfaces*, **5**, 1747–1756.

50. Xie, Y., Ou, Y., Ma, F. Y., Yang, Q., Tan, X. L., and Xie, S. H. (2013). Synthesis of multiferroic $Pb(Zr_{0.52}Ti_{0.48})O$-3-$CoFe_2O_4$ core-shell nanofibers by coaxial electrospinning. *Nanosci Nanotechnol Lett*, **5**, 546–551.

51. Yang, D. D., Xu, H. P., Wang, J. R., and Wu, Y. H. (2013). Nickel-multiwalled carbon nanotubes/poly(vinylidene fluoride) composites with high dielectric permittivity. *J Appl Polym Sci*, **130**, 3746–3752.

52. Yang, K., Huang, X. Y., Xie, L. Y., Wu, C., Jiang, P. K., and Tanaka, T. (2012). Core-shell structured polystyrene/$BaTiO_3$ hybrid nanodielectrics prepared by in situ RAFT polymerization: a route to high dielectric constant and low loss materials with weak frequency dependence. *Macromol Rapid Commun*, **33**, 1921–1926.

53. Yang, L., Qiu, J. H., Ji, H. L., Zhu, K. J., and Wang, J. (2014). Enhanced dielectric and ferroelectric properties induced by TiO_2@MWCNTs nanoparticles in flexible poly(vinylidene fluoride) composites. *Compos Part A: Appl Sci Manuf*, **65**, 125–134.

54. Yang, L. Y., Li, X. Y., Allahyarov, E., Taylor, P. L., Zhang, Q. M., and Zhu, L. (2013). Novel polymer ferroelectric behavior via crystal isomorphism and the nanoconfinement effect. *Polymer*, **54**, 1709–1728.

55. Yu, C. R., Wu, D. M., Liu, Y., Qiao, H., Yu, Z. Z., Dasari, A., Du, X. S., and Mai, Y. W. (2011). Electrical and dielectric properties of polypropylene nanocomposites based on carbon nanotubes and barium titanate nanoparticles. *Compos Sci Technol*, **71**, 1706–1712.

56. Yuan, J. K., Yao, S. H., Li, W. L., Sylvestre, A., and Bai, J. B. (2014). Vertically aligned carbon nanotube arrays on SiC microplatelets: a high figure-of-merit strategy for achieving large dielectric constant and low loss in polymer composites. *J Phys Chem C*, **118**, 22975–22983.

57. Zha, J. W., Dang, Z. M., Yang, T., Zhou, T., Song, H. T., and Li, S. T. (2012). Advanced dielectric properties of $BaTiO_3$/polyvinylidene-fluoride nanocomposites with sandwich multi-layer structure. *IEEE Trans Dielectr Electr Insul*, **19**, 1312–1317.

58. Zhang, L., Shan, X. B., Wu, P. X., and Cheng, Z. Y. (2012). Dielectric characteristics of $CaCu_3Ti_4O_{12}$/P(VDF-TrFE) nanocomposites. *Appl Phys A: Mater Sci Process*, **107**, 597–602.

59. Zhang, L., Yuan, S., Chen, S., Wang, D. R., Han, B. Z., and Dang, Z. M. (2015). Preparation and dielectric properties of core-shell structured Ag@polydopamine/poly(vinylidene fluoride) composites. *Compos Sci Technol*, **110**, 126–131.

60. Zhang, X., Shen, Y., Zhang, Q., Gu, L., Hu, Y., Du, J., Lin, Y., and Nan, C. W. (2015). Ultrahigh energy density of polymer nanocomposites containing $BaTiO_3$@TiO_2 nanofibers by atomic-scale interface engineering. *Adv Mater*, **27**, 819–824.

61. Zhang, X. H., Ma, Y. H., Zhao, C. W., and Yang, W. T. (2015). High dielectric performance composites with a hybrid $BaTiO_3$/graphene as filler and poly(vinylidene fluoride) as matrix. *ECS J Solid State Sci Technol*, **4**, N47–N54.

62. Zhang, X. J., Wang, G. S., Cao, W. Q., Wei, Y. Z., Liang, J. F., Guo, L., and Cao, M. S. (2014). Enhanced microwave absorption property of reduced graphene oxide (RGO)-$MnFe_2O_4$ nanocomposites and poly(vinylidene fluoride). *ACS Appl Mater Interfaces*, **6**, 7471–7478.

63. Zhang, Y., Wang, Y., Deng, Y., Li, M., and Bai, J. B. (2012). Enhanced dielectric properties of ferroelectric polymer composites induced by metal-semiconductor Zn-ZnO core-shell structure. *ACS Appl Mater Interfaces*, **4**, 65–68.

64. Zhang, Y. Y., Zhang, J. P., Gao, J. X., Li, H., Wang, E. P., and Hu, X. (2015). Effect of carbon nanotubes addition on properties of 0–3 pyroelectric ceramic/polymer composites. *J Electroceram*, **34**, 216–220.

65. Zhang, Z. C., Gu, Y. Z., Wang, S. K., Li, M., Bi, J. Y., and Zhang, Z. G. (2015). Enhancement of dielectric and electrical properties in BT/SiC/PVDF three-phase composite through microstructure tailoring. *Compos Part A: Appl Sci Manuf*, **74**, 88–95.

66. Zhou, W. Y., Chen, Q. G., Sui, X. Z., Dong, L. N., and Wang, Z. J. (2015). Enhanced thermal conductivity and dielectric properties of Al/beta-SiCw/PVDF composites. *Compos Part A: Appl Sci Manuf*, **71**, 184–191.

67. Zhou, Y., Wang, L., Ma, Y. H., Chen, D., Zhao, C. W., and Yang, W. T. (2015). Flexible /PET/BaTiO(3)/ layer-layer composite film with enhanced dielectric properties fabricated by highly loaded /BaTiO(3)/ coating with acrylic resin as binder. *J Appl Polym Sci*, **132**.

68. Zhou, Z., Carr, J., Mackey, M., Yin, K. Z., Schuele, D., Zhu, L., and Baer, E. (2013). Interphase/interface modification on the dielectric properties of polycarbonate/poly(vinylidene fluoride-co-hexafluoropropylene) multilayer films for high-energy density capacitors. *J Polym Sci, Part B: Polym Phys*, **51**, 978–991.

69. Zhu, J. M., Shen, J. B., Guo, S. Y., and Sue, H. J. (2015). Confined distribution of conductive particles in poly(vinylidene fluoride)-based multilayered dielectrics: toward high permittivity and breakdown strength. *Carbon*, **84**, 355–364.

Index

AC conductivity 23, 84, 102
actuated strain 150
AFM *see* atomic force microscopy
aluminum-coated silicon 123, 124
anion 19
aspect ratio 74, 79, 103, 140, 146, 147, 154
asymmetry 25
atomic force microscopy (AFM) 143
atom transfer radical polymerization (ATRP) 49, 53, 54, 57–59
atom 17–19, 71
ATRP *see* atom transfer radical polymerization

binary system 27
bio-assembled nanocomposites 141–143
bionanocomposites 139, 140, 142, 144, 146, 148, 150, 152, 154, 156, 158, 160, 162
biopolymers 139, 141, 159
bonded hybrid nanomaterials 175–177
breakdown strength 4, 53, 130, 176
BT nanoparticle 8, 173, 175, 176

calcination temperature 127
capacitance 21, 22, 153, 156
carbon nanomaterial 26, 27, 81, 108
carbon nanotube 5, 72, 79, 80, 151
  multiwalled 72, 75, 77, 151, 172,
  single-walled 72, 76, 78, 151
CCB *see* conductive carbon black
cellulose 146, 150
ceramic material 1, 35, 36, 49, 155
CFF *see* chicken feather fiber
chicken feather fiber (CFF) 145
clay nanomaterial 130
coating layer 52, 53, 55
comonomers 120
composite film 129, 150, 175
composite nanomaterial 184
composites 40, 74, 78, 140, 145–147, 149, 153, 154, 176–179, 182, 185
compression molding 99, 103, 120
concentration levels 39, 40
conch shell 141–143
conductive carbon black (CCB) 73, 75, 145
conductive nanomaterial 6, 8, 10, 26, 69, 70, 72, 73, 79, 151, 181
conductivity 102, 145, 147, 149, 152, 153, 159, 162, 183
copolymers 51, 53, 96, 112, 120–128, 130–134
copolymer structure 53, 96, 112, 120, 122, 124
core material 179, 181
coupling agent 44, 45
crystallinity 2–3, 100–104, 125, 127, 129
crystallization 105, 106

crystal morphologies 103, 122, 123
crystal 3, 105–109, 120, 123, 157
crystal structure 22, 35–37, 80, 95–96, 98, 100, 102, 108, 119–125, 128, 131

DEA *see* dielectric elastomer actuator
Debye model 24, 25
dielectric elastomer actuator (DEA) 149
dielectric loss 4, 10, 23, 39, 40, 43, 57, 142, 149, 156, 159, 172, 173, 179–181
dielectric loss factor 155, 160
dielectric nanocomposite 95, 96, 98, 100, 102, 104, 106, 108, 110, 112
dielectric properties 5–10, 21, 22, 26–29, 39–44, 50–55, 72–77, 79–81, 85–87, 96, 98–100, 102, 108, 125–129, 179–181
dielectric relaxation 17, 18, 20, 22, 24–26, 28, 30, 39, 42
dielectric relaxation model 24–25, 27
dielectric relaxation spectra 22, 25, 28
dielectric spectroscopy 152
differential scanning calorimetry (DSC) 101, 104–107, 109
diffraction peak 104, 107, 125
dipole moment 18–20, 35–38, 96, 98, 121
dipole orientation 121, 122
dipole 17–19, 23, 24, 28, 37, 119, 121, 122, 149
dispersion 6–9, 41, 44, 47, 50, 57, 74–75, 77, 79, 81, 86, 98–100, 171, 173, 176
dispersion nanoparticle 50
DSC *see* differential scanning calorimetry

EB *see* emeraldine base
effective medium theory 26, 27, 39, 102, 179
effective refractive index 161
elastic modulus 149, 150
elastomers 3, 4
electrical conduction 5, 6, 8, 24, 71, 111, 143
electrical conductivity 10, 43, 69, 99, 140, 145, 59, 183
electrical field 21, 76, 77, 80, 98, 121, 125
electron cloud 19, 83
emeraldine base (EB) 83, 84
energy storage 18, 96, 29, 30, 104, 119
energy storage application 10, 111, 112, 140, 156, 159, 162, 176, 181
energy storage capability 172, 181, 184
energy storage density 122, 123, 132, 181
energy storage performance 121, 128, 132, 173

fabrication 5, 6, 24, 42, 49, 51, 71, 72, 83, 96, 129, 151, 152
ferroelectric behavior 36, 38, 132, 133
ferroelectric crystal 121, 122
ferroelectric domain 121
ferroelectric properties 120
FET *see* field-effect transistor
fiber 145, 160–162
field-effect transistor (FET) 150, 151
fillers 27, 73, 109, 154, 180

GCN *see* graphitic carbon nanomaterial
gelatin 149, 159
geometry 22, 73, 74, 87
gold nanoparticles 160, 161
graphene 5, 71, 72, 85, 86, 154, 175
graphitic carbon nanomaterial (GCN) 5, 9, 71, 72, 76, 80, 83–85, 96

Havriliak–Negami model 24, 25
hybrid nanomaterials 49, 175, 176
hybrid nanomaterial systems 42, 172, 184
hybrid systems 172, 173
hydrogen bonds 147, 149, 150

insulator 4, 146, 152, 155, 181
interface 7, 8, 20, 21, 41, 54, 100, 121, 128, 149, 161, 176, 179, 181, 183
interfacial bonding 42, 47, 48, 50
interfacial polarization 18, 20, 21, 42, 45, 54, 57, 71, 72, 83, 86, 102, 145, 147, 179, 181, 184
interfacial structure 9, 36, 176, 180
ionic bond 18, 19
isolated nanomaterial 172, 173

keratin feather fiber (KFF) 145
KFF *see* keratin feather fiber

layer 38, 48, 72, 102, 129, 150
  passivation 48, 151
  thin dielectric polymer 153
leakage 10, 53, 71, 110
lightning rod effect (LRE) 143
localized surface plasmon resonance (LSPR)
low dielectric loss 38, 55, 95, 173, 174, 181, 182
LRE *see* lightning rod effect
LSPR *see* localized surface plasmon resonance

material 18–20, 22–23, 28–29, 36, 71–73, 132, 140, 156, 179
matrix 7, 8, 27, 55, 76, 80, 104, 129, 130, 158, 184
mechanical property 150, 152, 176
mechanism 22, 25, 26, 105, 109, 110, 123, 131, 173, 180
melt processing 99, 100
metal oxide 36, 156, 158
monomer 49, 124, 125, 132, 134, 154
morphology 2, 26–27, 36, 103, 104, 122–124, 157, 182
morphology factor 28

nanoclay 146
nanocomposite film 51, 103, 104, 106
nanofiber 5–7, 26, 74, 151, 181, 184
nanomaterial 3, 5–9, 26, 27, 42, 72–74, 79–81, 83, 107, 111, 112, 134, 171, 172, 175–177, 179, 180, 183, 184
nanoparticle 8, 36, 40, 42–48, 50, 52–55, 57, 58, 99, 102–104, 107–110, 127–128, 159, 161, 172, 176
nanoplate 69, 70
nanostructure 5, 7, 42, 43, 72–74, 128, 142, 172, 174, 181, 183, 184
  core–shell 24, 48, 179, 180
  hybrid 175, 182–184
nanotube 5, 7, 26
nucleation 100, 104, 107

pearl 142–144
percolation structure 80
permanent dipoles 17, 18, 20, 29, 37, 38
permittivity 22, 49, 76, 85, 147, 149
phase 3, 37, 98–100, 102–104, 106, 107, 109, 111, 112, 119, 121, 123, 126, 127, 131
  amorphous 20, 121
  crystalline 20
  polymer 80
phase crystal 104,110, 130
phase structure 22, 80, 104, 128, 129, 172
phase transformation 98–103, 105, 107, 109, 111, 120
phase transition 36–38, 40
PLA nanocomposites 85
polarity 50, 111, 147
polarization 17–19, 21, 29, 86, 124, 125, 142, 143, 159
  electronic 18, 19
  interfacial 69, 83, 84, 154, 182
  ionic 18–20
polarization mechanism 2, 18, 19, 22, 23, 25, 85, 181, 184
polar molecule 18, 20
poly(acrylic acid) 50
polyester 95, 96
polyester composite 145
polyetherimide 43, 74
polyethylene 80, 145
polymer blend 80
polymer chain 2, 50, 80, 134
polymer coating 49, 51, 53, 54, 57
polymer composite 35
polymer dielectric film 4, 120
polymer dielectric material 5
polymer material 1–5, 10, 22, 24, 25, 29, 30, 35, 50, 71, 80, 95, 102, 134, 182
polymer matrix 5–7, 9, 39, 40, 42–44, 47, 48, 50, 52, 54, 70, 71, 76, 78, 80, 102–104, 158, 176
polymer 1–4, 6–10, 24, 26, 27, 35, 36, 48, 54, 56, 58, 80, 102, 142, 147, 176, 179
  bio-based 154, 156
  semicrystalline 20, 96, 100, 104
polymer structure 2, 49, 54, 83
  hyperbranched 50, 54
  linear 50
polymer synthesis technique 53, 54
polymorphism 97, 124, 125, 128
polymorph 40, 134
poly(vinylidene fluoride) (PVDF) 4, 40, 51, 95, 96, 98, 100, 102, 104, 106, 108, 110, 112, 119
PVDF see poly(vinylidene fluoride)
PVDF matrix 8, 47, 48, 50, 69, 72, 102–104, 108, 172, 180, 184
PVDF nanocomposite 8, 47, 48, 50, 51, 70, 78, 79, 86, 96, 98–100, 106, 107, 109, 111, 125, 173, 179–181, 185

radical polymerization 53, 57
random copolymers 53
recrystallization 104, 106
relaxation time 24, 26, 102
renewable resources 140

SCF *see* short carbon fiber
semiconductor 71, 180, 181
SF *see* silk fibroin
shell layer 48, 49, 180
shell structure 48, 49
short carbon fiber (SCF) 73, 75
silane coupling agent 44, 85, 146
silk fibroin (SF) 151, 152

solvothermally reduced graphene (SRG) 86
SRG *see* solvothermally reduced graphene
starch 159
strain 4, 77, 79
stretched film 130, 131
stretching 3, 99, 104, 120, 122
  mechanical 76, 77, 80, 96, 106, 107, 120
  ratio 98, 99, 106
  temperature 99, 100
substrate 51, 120, 150
surface 7, 8, 48, 50, 53–55, 57, 58, 83, 84, 107, 108, 111, 147, 157, 179, 181
surface modification 8, 9, 36, 41–44, 46–50, 52, 57, 83–86, 100, 104, 107, 108, 176, 177
surface property 9, 44, 87, 107, 109, 111, 171
surface structure 83, 108
surfactant molecule 9, 48, 84
surfactant 8, 9, 43, 44, 47, 48, 49, 83, 85, 109, 111
synergetic effect 44, 104
synergy 80, 128, 134, 172, 181
synthesis 49, 53, 54, 58, 59, 96

temperature 19, 22, 36, 37, 96, 106, 120, 142, 144, 179
terpolymer 120, 132–134
tetrafluorophthalic acid 8, 47, 48
thermal property 4–6, 71
thermoplastics 3, 4, 99

uniform dispersion 6, 7, 47, 48, 50, 69, 74, 76, 173

variation 25, 142, 144, 147, 149, 154, 157, 160, 162, 172, 183
volume fraction 8, 26, 27, 40, 42, 102, 172, 173, 183, 185

An environmentally friendly book printed and bound in England by www.printondemand-worldwide.com

PEFC Certified

This product is from sustainably managed forests and controlled sources

www.pefc.org

PEFC/16-33-415

This book is made of chain-of-custody materials; FSC materials for the cover and PEFC materials for the text pages.